'Do you think you know everything about space? Even about the unconscious as space, and how it can be approached mathematically, topologically? Think again. This well-referenced book gives you food for thought, takes you by the hand to consider space anew – its dimensions, the importance of the hole, possible and impossible pathways, among other examples. This book does not just open your ears to new ideas but the space between them also.'
Vincent Dachy, *Psychoanalyst in the Lacanian orientation*

'*The Unconscious as Space* lucidly unfolds a compelling account of how thinking the Unconscious through mathematics, as space, affords those working with the unconscious, or the disjunctive fragments it presents us in conscious life, a coherent framing of why and how clinical interventions function. More than one thing can be true at the same time and this book has the maths to prove it! This readable book synthesises the arcane work of Lacan and other post-Freudians to date who attempted to think psychoanalysis through maths, making what might otherwise seem inaccessible and impossible both accessible and possible.'
Christopher Simpson, *Associate Lecturer, University of Northampton, UK*

'Lacan explored mathematics through his use of the Phi and -phi as representing the imaginary and symbolic phallus. In addition, he used the square root of –1 and imaginary numbers to represent the Real. The Real is outside the signifier although topology is inside letters. Freud explored psychic space and concluded that space is an extension of the psyche, "the Unconscious is outside or in the environment", as Lacan said. Not too many have followed Lacan in his use of mathematics, this book is an exception. The book covers both familiar and unknown territory. From Freud's unconscious of the not yet known, to the unknowable, and to Lacan's L'insu qui sait or unknown knowing, to Bion's grid, and to Matte Blanco's symmetrical unconscious. The book points to the next step regarding psychic space as quantum phenomena, in terms of contemporary physics, rather than the physics of Freud's time.'
Raul Moncayo, *Ph.D, Senior Lacanian Analyst, Chinese American Center for Freudian and Lacanian Analysis and Research, author of* Lacanian Psychoanalysis and American Literature *(Routledge)*

'This book offers what can be seen as a form of conversation between parts of psychoanalysis and parts of mathematics. Written with great clarity, it is a lively and serious venture into a world that is usually restricted to specialists. The author explores a terrain that is fundamentally concerned with human suffering, and she tries to seek out the underlying spaces that she holds to be present in any engagement with it. Alain Connes – an eminence in mathematics – has recently entered into a dialogue with a psychoanalyst colleague in France. Anca Carrington's text is a step towards the possibility of such a dialogue in Britain: not only here, but in places yet further afield.'
Bernard Burgoyne, *Emeritus Professor of Psychoanalysis, Middlesex University*

The Unconscious as Space

The Unconscious as Space explores the experience of being and the practice of psychoanalysis by thinking of the unconscious in mathematical terms.

Anca Carrington introduces mathematical models of space, from dimension theory to algebraic topology and knot theory, and considers their immediate psychoanalytic relevance. The hypothesis that the unconscious is structured like a space marked by impossibility is then examined. Carrington considers the clinical implications, with particular focus on the interplay between language and the unconscious as related topological spaces in which movement takes place along knot-like pathways.

The Unconscious as Space will be of appeal to psychotherapists, psychoanalysts and mental health professionals in practice and in training.

Anca Carrington is a London-based psychoanalyst with a particular interest in Lacanian analysis, and an associate member of the Centre for Freudian Analysis and Research. In her previous career she worked in spatial analysis and remains interested in the application of mathematical thinking to the field of psychoanalysis.

The Unconscious as Space

From Freud to Lacan, and Beyond

Anca Carrington

Taylor & Francis Group

LONDON AND NEW YORK

Designed cover image: Getty | NeoLeo

First published 2024
by Routledge
4 Park Square, Milton Park, Abingdon, Oxon OX14 4RN

and by Routledge
605 Third Avenue, New York, NY 10158

Routledge is an imprint of the Taylor & Francis Group, an informa business

© 2024 Anca Carrington

The right of Anca Carrington to be identified as author of this work has been asserted in accordance with sections 77 and 78 of the Copyright, Designs and Patents Act 1988.

All rights reserved. No part of this book may be reprinted or reproduced or utilised in any form or by any electronic, mechanical, or other means, now known or hereafter invented, including photocopying and recording, or in any information storage or retrieval system, without permission in writing from the publishers.

Trademark notice: Product or corporate names may be trademarks or registered trademarks, and are used only for identification and explanation without intent to infringe.

British Library Cataloguing-in-Publication Data
A catalogue record for this book is available from the British Library

ISBN: 9781032371085 (hbk)
ISBN: 9780367343538 (pbk)
ISBN: 9781003479284 (ebk)

DOI: 10.4324/9781003479284

Typeset in Times New Roman
by codeMantra

To Gary Cairns, for being

Contents

List of figures   xii
Preface   xiv
Acknowledgements   xv
List of abbreviations   xvi

PART I
Introduction   1

1 Introduction   3

2 The Freudian and post-Freudian unconscious as spatiality   8
    2.1 Freud's maps   12
    2.2 Post-Freudian emphasis   17

PART II
The unconscious as inaccessible space   35

3 The unconscious as infinity and the possibility of incompleteness   41
    3.1 Counting, recurrence and the spaces in between   42
    3.2 From numbers to posited dimensions   50
    3.3 Infinity and incompleteness in the unconscious   53
    3.4 The unconscious as inaccessible space between points of encounter   58

4 The rigour of spatial dimensions – of shadows and recurrences   63
    4.1 Flatland and beyond   66
    4.2 From higher to lower dimensions: projection   71
    4.3 From lower to higher dimensions: repetition   73
    4.4 Spatial dimensions in psychoanalysis   75

5 The unconscious as inaccessible and the exclusivity of the fourth dimension   79
  5.1 *Hyperspace – a primer*   *83*
  5.2 *The unconscious as four-dimensional space*   *87*
  5.3 *Dimensions in the clinic*   *90*
  5.4 *No dimensions and the inscription of impossibility*   *93*

PART III
The unconscious as domain of impossibility   99

6 Structures of the impossible   101
  6.1 *Mathematical representations of impossibility*   *102*
  6.2 *Holes and the unconscious*   *104*
  6.3 *The Oedipus complex as prohibition veiling impossibility*   *106*
  6.4 *Negations of impossibility*   *110*

7 The unconscious as topological space   115
  7.1 *Topology – a primer*   *116*
  7.2 *Holes and the unconscious, revisited*   *132*
  7.3 *Embedded or not – back to dimensions*   *134*

8 The unconscious as knots   140
  8.1 *Knots – a primer*   *141*
  8.2 *Knots as structure and pathways through language*   *145*
  8.3 *Knots in the clinic*   *146*
  8.4 *Knots and dimensions*   *148*

PART IV
Clinical implications   153

9 The spatial unconscious and the clinic of psychic structures   156
  9.1 *The recurrence of suffering*   *157*
  9.2 *Identification and subjectivity*   *160*
  9.3 *Dimensions in the clinic, revisited*   *161*

10	**Clinical implications**	165
	10.1 Topology in the clinic	*165*
	10.2 Clinical illustration	*171*
	10.3 Interpretation, the cut and the analytic act	*177*
11	**Concluding comments**	182
	Index	*187*

Figures

1.1	Known vs. unknown	4
2.1	The Freudian space of memory inscription	13
2.2	Topology of psychic agencies	14
2.3	Möbius band in fundamental polygon representation	15
2.4	Möbius band embedded in three dimensions	15
2.5	Freud's second topography	16
2.6	Freud's revised second topography	17
2.7	Knowledge in four dimensions	25
3.1	The number line	43
3.2	Number sets	49
3.3	The complex plane	51
3.4	Mapping of numbers in the complex plane	52
4.1	Sphere passing through linear space	69
4.2	Penrose triangle	70
4.3	Two-dimensional and unidimensional circle representation	75
5.1	Bijective mapping between segment and square	94
6.1	The multiplicative inverse function $f(x) = 1/x$	103
6.2	Jeu de Tacquin	106
7.1	Open space boundary	117
7.2	Möbius band in fundamental polygon representation	119
7.3	Embedded torus	120
7.4	Torus in fundamental polygon representation	121
7.5	Demand and desire on the torus	122
7.6	Klein bottle in fundamental polygon representation	123
7.7	Klein bottle as two Möbius bands	125
7.8	Klein bottle immersed in three-dimensional space	126
7.9	Projective plane in fundamental polygon representation	127
7.10	Decomposing the projective plane	129
7.11	Projective plane alternative representation	130
8.1	Reidemeister moves	144

9.1	Continuity in a higher dimension	159
10.1	Interlinked tori: desire and demand	169
10.2	Torus pathways (1)	174
10.3	Torus pathways (2)	176
10.4	Pivot signifiers	177

Preface

The story of this book begins with a question from Professor Burgoyne in a seminar organised by the Centre for Freudian Analysis and Research (CFAR) in London, in 2013, when he invited his audience to say something about using one of Lacan's schemas in a way that was different from what Lacan had intended. I never took up his challenge to submit an essay about it, but the question stayed with me; it made me think how, whenever encountering the Lacanian representation of the relationship between the signifier and the signified, I thought of arithmetic fractions. This led me to considering the relationship between numbers, and how that might differ from the relationship between words, and between words and meaning, which led me to thoughts about those numbers that can be obtained from operations applied to other numbers and those that cannot be created in that way. From the idea of numbers that exist but cannot be arrived at, I began to wonder about ordinary encounters with moments or experiences when we fail to explain through what we already know, about the structure that might generate such oddities and, beyond it, about that which remains unknown to us. The thought of a mathematical structure that is known to exist in the fourth dimension alone prompted me to consider how exploring the nature of such a space may tell us something about the nature of the unconscious, which is also a kind of space inescapably present, while remaining inaccessible as such.

What follows is what I uncovered in pursuit of this possibility and – as I write this – remains unknown and impossible to find in its entirety, while permissive of being uncovered further in the articulation of separate, future thoughts.

This pursuit of a possible answer aims for what can be grasped at the junction of psychoanalysis and mathematics, leaving aside the question of knowledge as addressed by philosophy.

London, 15 September 2023

Acknowledgements

In alphabetical order, Julie Burchill, Vincent Dachy, Astrid Gessert, Henrik Lynggaard, John Pate, Christopher Simpson – thank you all for asking me all the right questions.

Abbreviations

CFAR Centre for Freudian Analysis and Research
IPA   International Psychoanalytical Association

Part I

Introduction

Chapter 1

Introduction

The question I propose to address in this book is whether thinking of the unconscious in mathematical terms can shed new light on what remains unknown in our daily experience of being. Specifically, I am interested in exploring the unconscious as space, by addressing one specific question: can the articulation of unconscious as space, in a mathematical sense, help us know something about the nature of this most influential unknown in our lives?

In the psychoanalytic sense, the unconscious is a concept just over one century old. The question of how to decipher it has its own history, rooted as it is in Freud's efforts to convince his peers, and the world at large, that something called 'the unconscious' existed and that, although it was not knowable directly, one could know something about it and that one could think about it systematically.

In places and at times, this is an idea that remains difficult to accept for many. For those who accept it, much difficulty persists around clarifying what is actually meant by unconscious. While some think of it as an adjective, others use it as a noun. A number of questions persist, without a definite answer. Does this notion designate an attribute of a mental process, a place, a collection of elements with some common feature or a function? How is something unconscious encountered, and what can be known about what lies behind that encounter? Is the remainder unknown knowable? And what of the remainder beyond that?

Irrespective of how one might try to define this concept, one aspect of the unconscious is, however, essential: we are dealing with a particular kind of unknown, with something that can be known only indirectly, at best, and this is not without consequence. The prevailing relationship we have to the unknown, in general, is shaped by our relationship to what is known. With a keen eye on the recognition of what is not just unknown but unknowable, Magee (2016) stresses a common pitfall: 'We have a profound need, rooted in our need for survival, to believe that what exists does so in terms we can understand. The recognition that this is not so, and cannot be so, is disorienting' (p.65). We all rely on experience as our guide to knowledge, and psychoanalysts are no different. Freud developed his body of work on the basis of his clinical experience, and often referred back to the numerous cases on which he built his observations, and from which his theoretical formulations emerged. A devoted follower of Freud's, Jacques Lacan was at least as keen on experience as Freud, when it came to using it as test to the validity of his mostly theoretical and abstract thinking. Cléro (2002, p.29) observes that 'experience' is one of the terms that Lacan invokes the most in his seminars and writings, while theorising it the least.

Magee emphasises that the vast unknown is not just what is not known yet, but that which is not knowable at all with 'the apparatus we happen to possess, an apparatus that we may even possibly be' (2016, p.76), and stresses that 'most of reality is unknowable by us, and [...] unconceptualisable' (2016, p.85). So, in his view, unless we are one and the same with our experiences, whatever else we may be besides 'must remain forever unknowable' (p.117). The unknown is not just what we do not know yet, but most of all what we have no means of knowing and what we do not know that we do not know.

In contrast, Marcus du Sautoy, endlessly optimistic about the possibility of knowing, locates the Freudian unconscious in the partition of a space defined by the relationship between the known and the unknown, as 'unknown knowns' (Du Sautoy, 2016, p.11).

From a psychoanalytic perspective, the unknown knowns is the knowledge located in the unconscious, and it can be easiest understood as the repressed unconscious. But, as we are going to see, the unconscious is more complex than that. What can be mapped is not the true space of the mind, so to say, in the same way that the Tao that can be named is not the real Tao: 'Tao called Tao is not Tao' (Lao Tzu, 1993, p.1). Indeed, as Levinas (1999) emphasises, whilst knowledge takes up the datum, it also refuses it, as 'it aspires to riches beyond the frontiers' that are close to what is known (p.58).

When it comes to knowing, we have a wish to know not just what there is, but *where* it is. Producing the mapping depicted in Figure 1.1, as an illustration of du Sautoy's argument, was driven by precisely this wish to locate in order to comprehend.

Location is linked to existence, and thus to knowledge about that existence. It comes, therefore, as no surprise that addressing the question of psychic **space** is intrinsic to the development of psychoanalysis. Since his early work of 1895 on the *Project for a Scientific Psychology*, Freud grappled with the question of the

	known	unknown
known	Known knowns	Known unknowns
unknown	Unknown knowns	Unknown unknowns

Figure 1.1 Known vs. unknown.

mind in bodily terms, aiming to map what was where and thus echoing something of the older question of religion concerning the whereabouts of the soul. Freud was not content with his early theoretical project, which he abandoned in favour of the more immediate experience of the clinic. He soon came to recognise that the mystery of the psyche was not a question with an anatomical answer, and moved away from an idea of space as defined by biology, metrics and coordinates, to the consideration of relative positions described in functional terms.[1] Nevertheless, according to Cléro (2002, p.60), Freud remained haunted by the 'spatialisation of the spirit' until the end, with his last published words referring to it directly. A note from August 1938 (published posthumously) states, rather mysteriously: 'Space may be the projection of the extension of the psychical apparatus. No other derivation is probable. Instead of Kant's *a priori* determinants of our psychical apparatus. Psyche is extended; knows nothing about it' (Freud, 1941, p.300).

It has been argued that Freud was, unavoidably, constrained by the limits to the knowledge of his time. Yet, over a century after the publication of his *Project*, in the introduction to an interdisciplinary collection of papers on the question of reality and its dimensions, Cohen-Tannoudji and Noël (2003) stress the difficulty that persists in physics and beyond, in terms of apprehending, comprehending and representing what we call reality (p.9). Although we have learned much about the world in the last 100 years, the unknown persists in important and immediate ways.

Here, I want to concentrate on the way in which psychoanalysis relates to the unknown, in particular to the unknown of the individual unconscious, and on how a spatial approach might offer some fresh insight into thinking about this unknown.

When I refer to psychoanalysis, I have in mind both the body of knowledge it constitutes and the clinical practice it represents. Contemporary psychoanalysis covers a range of schools, all of which remain united by a number of shared assumptions, of which the coexistence of conscious and unconscious mental life is key, even though the way these relate to each other is conceptualised in less unified ways. The relationship between psychoanalysis and the unknown is shaped by the fact that psychoanalysts have concerned themselves with a particular kind of unknown, which they sought to address by developing an understanding of the origins and manifestations of human suffering. The question central to psychoanalysis was and remains aimed at what causes suffering in each and every one of us, and what determines the ways in which we respond to it.

As for space, I favour the definition given by words attributed to Newton[2] (1666): 'Spaces aren't themselves bodies; they are only the places in which bodies exist and move' (p.20), where a place is 'a part of space that something fills evenly' (p.1). This resonates with the relatively recent definition given by Alexandroff (1961), who designates as space that 'which appears as the place in which continuous processes occur' (p.1). Kasner and Newman (1940) emphasise our difficulty with talking about space and knowing what it is, given that we do not know what it would be like not to be in it (p.112). Yet, despite these limitations at the level of experience, mathematically, space is less difficult to comprehend.

The relationship to the unknown can be thought of in either deterministic or stochastic terms. Determinism is the belief in a cause, but a cause without intention, governed by the laws of chance (Charraud, 1997). How does this differ from pure randomness? In the former case, a given set of parameters will always lead to the same outcome, while in the latter the same sets of parameters can lead to unpredictably different outcomes. In this sense, in its fundamental assumptions, psychoanalysis is grounded in determinism, with early life experiences seen to be shaping to a large extent the ways in which each of us comes to encounter and respond to our own suffering. The possibility of change through the experience of analysis rests on the same grounds. Something new can be set in motion, without exactly aiming for or being able to anticipate a specific final outcome. The outcome of each and every analysis is not pre-determined by theory, but rests on each subject's possibility to create something new for themselves.

From the start, Freud conceived of the unconscious in spatial terms, as he attempted to map his thinking more than once throughout his work, never feeling quite satisfied with the outcome (1900, 1923). A small number of post-Freudian analysts expanded upon this spatial endeavour, openly calling upon mathematics as a guide (Bion, Matte Blanco and Lacan in particular). Starting from an overview of these contributions, I pursue further the relevance of rigorous spatial analysis to formalising thought about the Freudian unconscious in ways that have direct implications for the contemporary clinic. In particular, I explore two alternatives to the flat Euclidean space familiar to and employed by Freud: one is that of the unconscious as a Euclidean fourth dimension (in Part II) and the other is the unconscious as non-Euclidean topological space (in Part III), primarily in the sense that Lacan started to elaborate in his later work. In either case, the most basic notion of space as understood mathematically rests on the notion that a point is without dimension, a line has one dimension, a surface has two and a volume has three dimensions. It is important to note that none of these basic, widely used terms has a rigorous definition in mathematics, as they were left undefined in Euclidean geometry (Downing, 2009).

The remainder of the text is organised as follows. Chapter 2 concludes this first part, by offering an overview of Freud's development of the psychoanalytic concept of unconscious, and of its subsequent developments, both in the mainstream psychoanalytical literature and in the niche of psychoanalysis informed by mathematical thinking, with particular emphasis on spatial aspects. Part II considers the unconscious as space defined by dimensions, starting with an exploration of the number space (Chapter 3), through the geometry of three-dimensional space (Chapter 4), and into the fourth dimension (Chapter 5). Part III examines relevant insights from topology, introducing the idea of a formal representation of impossibility (Chapter 6), considering the specifics of topological spaces (Chapter 7) and of knot theory (Chapter 8). Throughout, the mathematical concepts considered are introduced and examined from the perspective of their relevance to psychoanalytic thinking. Part IV is devoted to a closer exploration of the clinical implications of the mathematical insights considered, both at the level of theoretical formulation (Chapter 9) and in the realm of the daily clinical encounters (Chapter 10). Chapter 11 brings together some concluding comments.

In order to get the most out of this exposition, my suggestion to the reader is to treat it as an old-fashioned sort of book, where chapters are to be read in the order in which they are presented.

In essence, the main thesis put forward here is that the unconscious is structured like a space, operating in ways defining of the interplay between spatial relations. Approaching it as a spatial structure yields new insights into the working of the mind and of what is at stake in our clinical work.

Notes

1 This shift echoes the move in the mathematical understanding of space away from metrics (Euclidean space) to relative positions (topological space), an idea to which we will return more than once in what follows.
2 The text attributed to Newton is the work of philosopher Jonathan Bennett, who undertook the task of making texts of early modern philosophers more accessible to a contemporary readership.

References

Alexandroff, P. (1961) *Elementary concepts of topology*. New York: Dover Publications.
Charraud, N. (1997) *Lacan et les mathématiques*. Paris: Anthropps.
Cléro, JP. (2002) *Le vocabulaire de Lacan*. Paris: Ellipses.
Cohen-Tannoudji, G. and Noël, È. (eds.) (2003) *Le réel et ses dimensions*. Editeur: EDP Sciences.
Downing, D. (2009) *Dictionary of mathematics terms*, 3rd ed. New York: Barron's Educational Series.
Du Sautoy, M. (2016) *What we cannot know*. London: 4th Estate.
Freud, S. (1900) The interpretation of dreams. SE4 & SE5.
Freud, S. (1923) The ego and the id. SE19, pp.3–66.
Freud, S. (1941 [1938]) Findings, ideas, problems. SE23, pp.299–300.
Kasner, E. and Newman, J. (1940) *Mathematics and the imagination*. New York: Dover Publications.
Lao Tzu (1993[c.605-c531BC]) *Tao Te Ching*. Translated by Stephen Addis and Stanley Lombardo Cambridge: Hackett.
Levinas, E. (1999) *Alterity and transcendence*. London: Athlone Press.
Magee, B. (2016) *Ultimate questions*. Oxford: Princeton University Press.
Newton, I. (1666) *Descartes, space and body*. https://www.earlymoderntexts.com/assets/pdfs/newton1666.pdf

Chapter 2

The Freudian and post-Freudian unconscious as spatiality

Freud's formulations around the notion of the unconscious, which constituted the core of his theoretical and clinical work, underwent development throughout his life. What is of particular interest here is the way in which formal elements of thinking about the psyche in spatial terms underpinned the evolution of this Freudian concept. In essence, Freud abandoned early attempts to identify specific locations for mental processes in favour of considering the relative positions of the agencies involved in registering and processing the experience of being.

Despite the importance of these spatial foundations, only a small number of subsequent analysts engaged explicitly with the question of the relevance of space to understanding the nature and operation of the unconscious, and the particular kind of knowledge that it constitutes. Up to this point, these contributions have not come together in a cohesive body of work.

The Freudian unconscious is an 'elsewhere' of a particular kind, a spatial unknown that cannot be accessed directly or even apprehended in its totality. This view changes the common notions of 'internal' and 'external' and offers a new basis for examining the question of the dis/continuity to which the concept of unconscious itself was introduced by Freud as a necessary hypothesis.

The Freudian unconscious is a subject on which much has been written, primarily in the context of post-Freudian developments that focus on the ego (e.g. Akhtar and O'Neil, 2013). Given that so much has been expanded upon in this direction, I restrict my emphasis here on a particular aspect, namely, on Freud's specific efforts to formulate his thinking in a spatial manner.

It is often said that Freud was the one to 'discover' the unconscious. That is not to say that before him the idea of something escaping consciousness was not around (see, e.g., Grose, 2014). What is uniquely Freudian about the unconscious, such as we think of it in psychoanalysis, is how Freud conceptualised this as a system, in a way that enabled him to develop at the same time both a theoretical way of understanding the workings of the mind and a clinical technique. Starting with him, the territory where an answer to the question of human suffering and the workings of the psyche could be sought became narrowed down to the conceptual space of the unconscious. In other words, the unconscious became recognised as the domain of the influential unknown, an unknown that puts its rigorous – if mysterious – mark over the human experience of finding one's place in the world.

Initially, Freud distinguished between the unconscious as an adjective describing those mental processes not subject to consciousness and the unconscious as a noun, by which he designated a radically separate domain of the psyche. This was not to be understood as merely one opposed to consciousnesses, but radically different from it. He later also emphasised the dynamic view of the unconscious, by which he designated those latent ideas which remain kept 'apart from consciousness in spite of their intensity and activity' (Freud, 1912, p.262), and which generate emotional conflict and therefore suffering. In the same paper, which Freud wrote and delivered in English, he concluded that 'The Unconscious' as a system is the most significant sense this term has in psychoanalysis (*ibid*, p.266).

The first stage in Freud's theoretical formulation of the unconscious was that of the affect-trauma model, lasting from the mid-1880s until the late 1890s (Sandler et al., 1997). During this period Freud came to see that the division between the conscious and unconscious parts of the mind was universal, and not just specific to neurotic patients, although it was during his work with them that he formulated his understanding. Specifically, he examined the partitioning between conscious and unconscious registration through his explorations of repression (Freud, 1915a), as he identified and tracked two different component representations of an instinct (drive[1]), namely, an unacceptable idea, which becomes repressed, and a corresponding quota of affect, which becomes separated from the idea in question and follows a trajectory of its own. Thus, the role of repression is that of avoiding unpleasure, either by keeping the displeasing idea unconscious or by pushing it out of consciousness. As for the affect, Freud proposed that when the quantity of affective energy became too large to be handled by the conscious mind, the surplus was repressed, that is to say, forced into what he called the unconscious mind. This surplus affective energy was that of emotions associated with traumatic ideas or memories which neurotic patients found incompatible with their normal standards of morality and conduct and could therefore not be absorbed or discharged in a normal way, thus leading to the formation of symptoms.

In a sense, this formulation was implicitly spatial, as it explained a psychic process in economic terms, based on a hydraulic view of quantities moving from one location to another. However, Freud's preliminary topographical conceptualisation of the mind was not outlined explicitly in spatial terms until *The Interpretation of Dreams*, in 1900, where, in Chapter 7, he sketched the distinction between conscious and unconscious mind (Freud, 1900). In its attempt to link the psychic process to the physical structure and anatomy, this model contains echoes of the earlier *Project for a Scientific Psychology* of 1895, which was very much conceived with the aim of uncovering the physical habitat of psychic processes (Wollheim, 1991, pp.43–44). The question of location was central to this endeavour. In this early topography, Freud introduced the idea of psychological systems linked spatially along the (vertical) axis of depth, distinguishing between conscious and unconscious aspects of the mind as layers where psychoanalysis operated as a kind of archaeological dig in search of the repressed.

However, it was not until some 15 years later that the topographical model proper was presented in detail in *The Unconscious* (Freud, 1915b). This model, built on the earlier ideas, distinguished between the Unconscious (*Ucs*), Preconscious (*Pcs*) and Conscious (*Cs*), domains conceived of as distinct systems characterised by their own functions and laws, rather than just as attributes of the contents of the mind or degrees of consciousness. This became known as Freud's first topography.

The second topography, proposed some eight years later, differentiated the agencies of the id, ego and super-ego, each with their own unconscious component (Freud, 1923b). Laplanche and Pontalis emphasise that these various subsystems or agencies are demarcated in such a way that 'they may be treated, metaphorically speaking, as points in a psychic space which is susceptible to figurative representation' (Laplanche and Pontalis, 1988, p.449).

According to the first topography, the three systems (*Ucs*, *Pcs*, *Cs*) are separated by boundaries of censorship which work to inhibit movement from one system to another. Defences are in place to prevent unconscious material from emerging into consciousness or, failing that, at least to modify any material that threatens to escape. The movement in the opposite direction, away from consciousness, is one of repression (e.g. 'forgetting'), one of regression (e.g. dreaming) or indeed the 'excavating' work of psychoanalysis, which is often met with resistance.

This first topography was contemporary with Freud's first theory of the drives and the foundations offered by the pleasure principle. The basis of his libido theory at that time was the opposition between ego (or self-preservative) forces and sexual (or reproductive) ones, as a psychological reflection of 'the biological fact that the living individual organism is at the command of two intentions, self-preservation and the preservation of the species' (Freud, 1933, p.95).

More specifically, mental functioning according to the pleasure-unpleasure principle was seen to ensure that pleasure is maximised and displeasure kept to a minimum. In other words, Freud postulated that the psychical apparatus was regulated by the avoidance or discharge of unpleasurable tension (Laplanche and Pontalis, 1988, p.322). To achieve this, either activity is increased in the pursuit of relief or defence mechanisms are introduced to reduce the tension. As in the affect-trauma model highlighted earlier, repression retained a central role. The three systems (*Ucs*, *Pcs*, *Cs*) were seen as separated by censorships set to inhibit and control the movement of psychic material between systems. The newly added system *Pcs* was allocated a double role, namely, that of functioning as 'reservoir of accessible thoughts and memories' and as 'censor capable of modifying instinctual wishes of the system unconscious and so render them acceptable to the system conscious' (Bateman and Holmes, 1995, p.32). Within this model, a permanent pressure is maintained in the unconscious, by sexual instincts or drives, necessitating a continuing contrary, resisting force to prevent them from reaching consciousness. Resistance is also present when the journey in reverse is attempted in analysis, from the conscious to the unconscious mind. Compromise formation between the repressed sexual instincts and the repressing ego instincts makes it possible for suitably disguised repressed ideas to reach consciousness in the form of symptoms (Freud, 1923a, p.247). On the basis of this understanding of psychic processes, clinically, Freud did not set out to eliminate symptoms. Instead, he took them seriously as manifestations of the unconscious that not only expressed suffering, but also held, in some encrypted way, the best clues to the possibility of change.

The topographical model was accompanied by and built upon Freud's distinction between primary and secondary mental processes, with the former characteristic of unconscious mental activity and the latter of conscious thinking (Freud, 1911).[2] Primary mental processes – typical of dreams, phantasy and infantile life – are governed by the pleasure principle, reducing the unpleasure of instinctual tension by hallucinatory wish fulfilment (Rycroft, 1972). These processes do not take account of the laws of time and space; one symbol can carry several meanings, and one meaning can be carried by several symbols. Secondary mental processes – characterised by rational thinking and the rule of ordinary laws of logic – are governed by the reality principle, whereby unpleasure associated with instinctual tension is reduced through adaptive behaviour.

That which constitutes the Freudian system unconscious cannot be directly apprehended; at best, something about it can be gleaned through what lurks at the periphery of experience, and is usually dismissed as unimportant or erroneous, that is to say, through unconscious formations that escape censorship: dreams, slips of the tongue, bungled actions, and the like. The unconscious in this sense contains repressed memories, those of childhood as well as later ones, alongside representations of drives and impulses. While it was clear to Freud that all repressed material was unconscious, he also understood that not all that was unconscious was the repressed (the unrepressed unconscious). Later, this came to be known, through Lacan's work, as the real unconscious (Soler, 2014). Also, Freud's clinical investigations showed that conscious awareness of repression did not put an end to the patient's problems. In other words, repression could be both conscious and unconscious at the same time. That is to say, mental functioning was not ruled by symmetry or easily submitted to neat, quasi-anatomical partitioning.

Much of the work of psychoanalysis is built on the understanding that the cause of suffering lies in the tension between repressed ideas and impulses seeking expression and satisfaction, on the one hand, and the opposition to their manifestation imposed by their unacceptable nature. What makes them unacceptable is their incompatibility with one's own carefully (and unconsciously) constructed ideal image of oneself which, in turn, is shaped by subtle unconscious interactions with one's world and phantasy about one's place it it. As Grose (2014) sharply puts it: 'For Freud, the unconscious isn't a big mess. It's a system. The reason that the things in it have to be kept away from consciousness is because they're deemed unsettling to the person's good image of themselves' (p.13). In other words, we suffer because we find it very hard to keep our wishes somewhere else, some place other than the one where we try to lead acceptable, orderly lives. In this sense, suffering is a failing in an effort to impose a kind of spatial partitioning of the experience of being.

After a consideration of Freud's key formulations of the unconscious in spatial terms, we turn to an overview of the main post-Freudian developments stemming from these, as far as the unconscious as knowledge, and then as space, is concerned.

2.1 Freud's maps

Laplanche and Pontalis (1988) emphasise as particularly Freudian the topographical and dynamic aspects of the unconscious as noun, rather than as adjective (pp.474–475). In particular, they draw attention to the unconscious as a system, 'as a particular "psychical locality" that must be pictured not as a second consciousness but as a system with its own contents, mechanisms and – perhaps – a specific "energy"' (p.475).

For Freud, in the first instance, the existence of the unconscious was something he needed to *assume* in order to be able to 'develop the theory of dreams further or to interpret the material met with in dream-analyses' (Freud, 1905a, p.161). He made it clear that 'the unconscious is something which we really do not know, but which we are obliged by compelling inferences to supply' (p.162). In this sense, the unconscious was not so much 'discovered' as it was posited as something that would make it possible to make some sense of human suffering. Its life began as a name for the gaps and discontinuities which insisted to disrupt the attempted order that the conscious mind has always tried to impose upon our lived experience.

Also, to start with, for Freud, the unknown that the unconscious represented was largely hidden, but knowable – albeit indirectly – something that could be approached with the methodical determination of an archaeological dig (e.g. Freud, 1896). This is, essentially, a spatial approach, in that it locates the hidden unconscious 'elsewhere'.

In his later work, Freud shared an increasing awareness of an endlessness inherent in the analytic process which decades earlier he had viewed with the optimism of the scientist who regards the unknown as that which is not known yet (e.g. Du Sautoy, 2016). Indeed, some four decades after his early use of the archaeological metaphor, Freud placed his emphasis not on the possibility of a successful uncovering of something hidden, but on the limitations of psychoanalysis, on its difficulties and inherent obstacles (Freud, 1937). The language in which he described these difficulties retained spatial undertones, as he wrote of the problems encountered in terms of degrees of *mobility* or *plasticity* of the libido (pp.241–242 – my emphasis). As for the endlessness of analytic work, the original *Unendliche* translated as interminable in the Standard Edition, which hints at a time aspect in English, also indicated a spatial aspect of endlessness in the original German.

Freud attributed a significant share of the opposition to psychic change – and hence of the limits to what is possible in clinical work – to the death drive. Its spatial equivalent was not something he elaborated explicitly, although he did express the implicit spatial conundrum in the title of his most influential paper on this: the answer to repetition compulsion lay *beyond* the pleasure principle (Freud, 1920).

A post-Freudian of Lacanian orientation, Jean-Gérard Bursztein (2016) argues that Freud was thinking topologically, even though he did not formulate his theories in those terms explicitly, influenced as he was by the Galilean epistemology prevailing at the time rather than the later Einsteinian epistemology that is not defined by forces but by space (p.xix).

It could be argued that Freud's inclination to map out the relative positions of the elements he identified as agencies of the psyche, the maps of which he proposed as topographies, was rooted in the neurological tradition of his time, which aimed to localise functions or ideas in various parts of the cortex. Traces of this way of thinking are apparent in his *Project for a Scientific Psychology* of 1895, where functional and anatomical considerations intermingle. Yet it is this very same tradition which Freud increasingly criticised and against which he developed his own models of the mind, with emphasis placed increasingly on function rather than on anatomy.

As early as 1895, in *Studies on Hysteria*, Freud described the unconscious in a spatial configuration, with an organisation of concentric layers of memories around a pathogenic nucleus (Freud, 1895, p.289). This turned the analytic treatment into a journey towards this nucleus, confronted by increasing resistance, a representation resonating with his view of the analytic process akin to an archaeological dig.

An early representation of his evolving spatial conception emerged in a letter of 1896 to Fliess, where he considered the layering of memories in multiple inscriptions. There, he sketched the relative positions of perception and unconsciousness as given in Figure 2.1.

The movement he described here is from neurones specialised in perception and devoid of memory (W^3), through a first registration of the perception (Wz^4), on to what he called here unconsciousness (Ub^5), and where an inscription of conceptual memories that are not accessible to consciousness occurs, then to a preconscious registration (Vb^6), where a third transcription, attached to word presentations, occurs, and finally to consciousness ($Bews^7$). The quality of this partitioning appears rather homogeneous, with all the points of registration forming a sequence along one line.

However, only five years later, in *The Interpretation of Dreams* of 1900, Freud began to formulate a more complex spatial understanding, with the unconscious recognised as a radically different kind of space, which is not simply further along or deeper relative to consciousness, but of an entirely different kind. In a different dimension, one could say. At this point, he takes on an idea from G.T. Fechner, whom he credits for the recognition that 'the scene of action of dreams is different from that of waking ideational life. [...] This is the only hypothesis that makes the special peculiarities of dream-life intelligible' (Freud, 1900, p.536). The scene upon which dreams are enacted is not a mere extension of waking mental

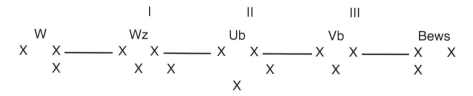

Figure 2.1 The Freudian space of memory inscription.

functioning, but rather a scene of another kind, a genuinely 'other scene' (*eine andere Schauplatz*). In other words, another kind of space.

Freud was explicit about his search being one for a 'psychic locality' (Freud, 1900, p.536), which he recognised not as an anatomical one, but more as something analogous to the formation of images in an optical apparatus, a kind of virtual location, akin to 'regions in which no tangible component of the apparatus is situated' (Freud, 1900, p.536).

Thus, the mental apparatus as a compound instrument formed of the agencies, instances or systems that Freud identified was conceived of as ordered either in space or in time. If we take time as representing a fourth dimension, we can see that it is a kind of hyperspace that Freud is trying to construct. In topology, a hyperspace is a topological space within which some of its elements form another topological space. The illustration he gives is in fact defined not by dimensions in Euclidean space, but by surface topology. More on this in Part III.

Indeed, a close examination of Freud's first topography as depicted in 1900 shows that what he proposed there was an arrangement which situated the unconscious relative to perception and to consciousness in a relationship of continuity, such that the elements of the system appeared as both distinct and inseparable. In this representation, psychical processes advance from the perception end (*Pcpt*) to the motor end (*M*). The perception was represented as specialised in receiving perceptual stimuli, of which it retains no trace, while a second system '*behind* it [...] transforms the momentary excitations of the first system into permanent traces' (Freud, 1900, p.538; my emphasis). These traces retain more than the mere content of the perceptions, and each excitation leaves 'a variety of different permanent records', labelled *Mnem* (mnemic system), and which are 'in themselves unconscious' (Freud, 1900, p.539). These unconscious memories have no access to consciousness except via the preconscious, and are therefore subject to modification.

In a footnote added in 1919, Freud addressed being confronted with the mysterious nature of this space: 'If we attempted to proceed further with this schematic picture, in which systems are set out in linear succession, we should have to reckon with the fact that the system next beyond the *Pcs* is the one to which consciousness must be ascribed – in other words that *Pcpt* = *Cs*' (Freud, 1900, p.541). In other words, *Cs* is at both ends of this representation – see Figure 2.2.

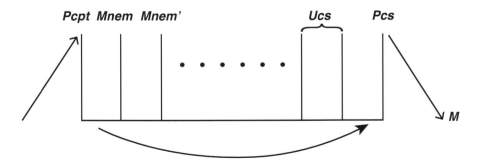

Figure 2.2 Topology of psychic agencies.

What Freud did not have at the time were the means to conceptualise the fact the he was depicting the relative positions of the agencies of the psyche as a continuous space with a twist, which makes his diagram (Figure 2.2) equivalent to a Möbius band in fundamental polygon representation with directed edges (Figure 2.3).

Figure 2.3 Möbius band in fundamental polygon representation.

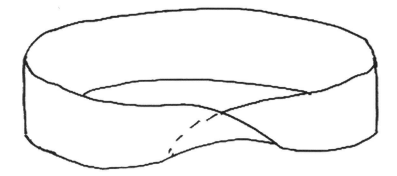

Figure 2.4 Möbius band embedded in three dimensions.

The Möbius band is a topological surface that has two sides at any one point (locally), but is one single surface overall, with a continuous path crossing both apparent sides. In the fundamental polygon representation (Figure 2.3), this appears as a flat surface with two opposed sides 'zipped' together in the direction indicated by the arrows (i.e. the top left corner joins up with the bottom right corner, and so on along the sides with arrows, until the bottom left corner joins the right top corner). A more familiar representation is the way this particular space is as embedded in three dimensions (Wells, 1991, p.152), like in Figure 2.4.

Without making this particular link explicitly, it was Lacan (1953), decades later, who made topology central to psychoanalytic thinking, using the Möbius band to make it clear that the unconscious is not some deeply hidden place, but a spatial structure that does not rely on a distinction between internal and external or between surface and depth.

16 Introduction

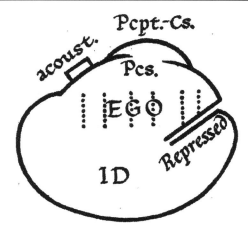

Figure 2.5 Freud's second topography.

This resonates with and builds upon Freud's consideration of the relationship between internal and external perceptions and the difficulty with separating these in a radical way, in his second topography (Freud, 1923b, p.21). The diagram Freud presented there proposed a new mapping of the psyche, where *Pcpt* and *Pcs*, which were at opposite ends of a linear representation of 1900, appear continuous with each other, yet on either side of a dividing line – see Figure 2.5.

This representation is also consistent with the Möbius band, where continuity is also a discontinuity, at the twist. Yet this twist is not at one point alone; it is everywhere, something that can be more accurately seen in the fundamental polygon representation than in the version embedded in three dimensions (i.e. Figure 2.4 vs. Figure 2.3). Freud is explicit on this point, as he objects to the use of sharp frontiers, akin to those used in political geography, proposing instead a representation 'by areas of colour melting into one another as they are presented by modern artists. After making the separation we must allow what we have separated to merge together once more' (Freud, 1933, p.79). Figure 2.5 is a depiction of such a space, hinting at both continuity and separation.

In a new representation which Freud included in his introductory lecture XXXI on *The Dissection of the Psychical Personality*, he added the super-ego to the map, as a distinct agency – see Figure 2.6. Also, at this point he moved away[8] from using the abbreviation *Ucs* as a way of merely designating a particular 'mental province' (p.71).

One could argue that Freud grappled with the spatial attributes of the unconscious throughout his work, without quite arriving at a satisfactory formulation. Like everything else, the spatial approach to the psyche is open to misunderstandings, with the unconscious in danger of being thought of as a kind of spare room into which unwanted things go and from which sometimes they also come out. In a classic IPA (International Psychoanalytical Association) text on technique, for

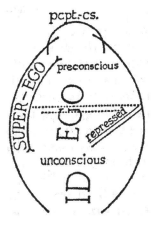

Figure 2.6 Freud's revised second topography.

instance, Etchegoyen (1999) refers to Freud's first topography as 'the model of the three boxes – that is, the three systems: unconscious, preconscious, and conscious' (p.109). Such oversimplification remains on the level of the anatomical view from which Freud had started, but from which he had distanced himself in favour of a much more nuanced, if more difficult to represent and comprehend visually, spatial conception of psychical processes.

In 1915, Freud proposed the hypothesis of the unconscious as a way of gaining meaning, as a necessary interpolation between observable conscious acts that would otherwise remain disconnected and unintelligible (Freud, 1915b, p.167). In this sense, the unconscious becomes that which introduces continuity in the apparent discontinuity of the phenomena of mental life. This is also an understanding of psychic life that implies a space without holes, a continuous domain, the whole of which is not accessible as such, but can only be experienced in fragmented or discontinuous ways.

Something that exists but cannot be perceived in its totality can be further understood mathematically in at least two ways: in terms of dimensions or in terms of topology. I explore each of these in turn in Parts II and III, respectively.

2.2 Post-Freudian emphasis

The way in which psychoanalysis evolved after Freud is usually considered in terms of the focus of the prevailing schools that developed starting from the first part of the last century: ego psychology in the US, object relations in the UK, Lacanian on the continent (interestingly identified by the name of the leading analyst whose teachings are followed, rather than by its explicit emphasis on language and speech, as well as on topology).

It can be argued that, irrespective of school, post-Freudian analysts developed their own views about the unconscious, primarily in terms of knowledge – what can be known, how and what is the place of the unknown in the experience of being. Some of them took a mathematical approach to this question, but only a small number of analysts pursued spatial considerations explicitly.

Here I would like to concentrate briefly on those approaches that focused on the unconscious as knowledge while taking some interest in a mathematical approach, and then expand in more detail on those approaches that proposed a spatial view of the unconscious, also grounded in mathematical underpinnings.

In a review for what at the time was a new play in London, on the theme of mathematics, the *New Scientist* remarks that 'mathematics … is not really our friend… it goes where we can't' (New Scientist, 2016). Yet precisely because of that, mathematics manages to bring order to our knowing, offering an appealing formalism that fits our experience of the world in ways that appear to be custom-made at times. As Maimonides (1190) put it, mathematics is a reservoir of such 'pre-adapted' abstract forms – and that is what its formalism is about (cited in Teissier, 1997). Mathematics might not be a friend, but it certainly is a reliable companion to thinking about what we can experience, and a much needed envoy to that part of the unknown where we cannot accompany it in the sense of experience as unmediated knowledge. Ignacio Matte Blanco, one of the main psychoanalysts engaged in the pursuit of the relevance of mathematics to psychoanalysis, draws attention to the emphasis that Braithwaite, the philosopher of science, placed on mathematics as providing 'a variety of methods for arranging hypotheses in a system' (Braithwaite, 1953, cited in Matte Blanco, 1975, p.7). Along the same lines, Bursztein (2017) goes further by proposing that mathematical formalisations are not to be regarded as mere abstractions, but as ways of engaging with the structure of the unconscious, as modes of accessing it (p.9). His view rests on that of Lacan, who insisted that topology is not to be taken as a metaphor, but studied as the very structure of the psychic space (Lacan, 1973).

It is no surprise, therefore, that a number of attempts at linking mathematics to psychoanalysis have been made so far, in the pursuit of introducing some order to the kind of unknown that the unconscious confronts us with on the level of experience. It is notable that most such attempts have remained at the stage of speculation, and that they seem to have evolved in isolation from each other, with various analysts reaching out for various elements of mathematics, in a rather fragmented way. Often, such attempts have not been developed through to their full consequences, either for theory or for the clinic, but rather were left as open invitations for others to pursue and bring about a kind of rigour that remains both promised and elusive. There are many such starting points, but not many analysts have completed their intended journeys into the full exploration of what appears to be the promise of a fruitful synergy between psychoanalysis and mathematics. Some notable exceptions in terms of spelling out in some detail what the interconnections between the two disciplines look like exist, nevertheless, in particular in the works of Wilfred Ruprecht Bion, Imre Hermann, Ignacio Matte Blanco and Jacques Lacan.

2.2.1 The unconscious as knowledge

Freud understood human suffering as the result of the interplay between opposing forces, which he regarded as ultimately grounded in instinctual[9] dualism (Freud, 1905b, 1920). Motivated by an interest in this dynamic view of the unconscious, a number of proposals have been made to pursue the linking of psychoanalysis and mathematics in terms of modelling dynamic processes or chaos, partly so as to mimic ways in which mathematics has helped model biological processes (see, e.g., Langs, 1988), or to expand the insights into non-linear processes to explore human development and therapeutic interactions (see Rose and Shulman, 2016, for an overview). The primary interest of such undertakings was in considering the type of models that calculus – the mathematics of change – could offer, in order to 'discover the underlying order beneath the surface chaos' (Langs, 1988, p.206). Although some of these models thus put forward refer to topology (and hence to space), little explicit links to the psyche or to analytical work of the clinic are made, beyond the suggestion that such models might illuminate beyond the reach of ordinary theorising in psychoanalysis.

Bion, often regarded as one of the more mathematical followers of Freud, and a member of the British object relations group, focused primarily on ways of organising knowledge rather than on the unknown, although he emphasised its central place in the analytic experience. He distinguished between two kinds of knowledge, which he designated as K and -K, as a way of contrasting knowledge that is based in experience (K) with cognition driven by reason, negatively charged in his view (-K) (Bion, 1962). His most formalised contribution in this regard is the grid he introduced in 1963 and elaborated subsequently, where he mapped thoughts and communications according to their degree of complexity and according to their functions (Bion, 1963).

Thus, in terms of complexity, Bion proposes that thoughts range from so-called β-elements ('undigested' raw sense-data, undifferentiated from things), through α-elements (processed β-elements), dreams and myths, to concepts, scientific deductive systems, all the way to what he regards as the most abstract – algebraic calculus. The relation between α and β is very much the relation between a function[10] and its argument in mathematics, between y and x in the familiar expression $y = f(x)$. In terms of functions, communications in the session range from what he calls definitory hypotheses (e.g. telling a patient that what she/he is experiencing is depression – Bion, 1963, p.18), summaries (which he calls *notation*), pointing out particular elements (which he calls *attention*), to enquiry and action (by which he means interpretations that constitute a communication that 'will enable the patient to effect solutions of his problems of development', Bion, 1963, p.20).

All these categories are coded according to numbers that identify columns, and letters that identify rows of a grid. Thus, a session can involve combining an E1 with a D1 to produce an F1, or a D1 can develop into an E1, etc. His aim was to produce a tool that could generate a mathematical model of the analytic interaction, moving away from the content of speech towards its functions. Like Lacan's mathemes, these coded elements were meant to formalise something in the communication among analysts and in the transmission of psychoanalysis. Bion aimed for this grid to be used as an instrument for scanning of material in sessions, that is to say, both in the patient's and in the analyst's speech, and of the analyst's thought. Movements across the grid were to operate as indicators of transference and of change in treatment (Bion, 1963). Spatiality is present here only inasmuch as Bion located various moments of speech in the analytic sessions along the grid coordinates he proposed, as if on a map, on which some relative positions could be captured.

Bria (1981) singles out Bion's notion of transformation in the sense of spatial mapping, and links it to Matte Blanco's understanding of the mind in spatial terms. In both cases, there is an explicit recognition that ultimate reality is unknowable in itself and that the phenomenology of our mental world cannot be adequately represented in terms of the common geometry used to represent the structure of physical space (p.505). In this sense, patients arrive with certain images of the world, which are transformations of the real with which they are not in a univocal relationship. The analytic interpretation works on further transforming this image, operating as transformations of transformations which originates in a 'fact' which can be symbolised in infinitely many ways (Bria, 1981, pp.506–507). The source of these 'facts' is what Bion designates as O, the spatial origin of knowledge – read by some as the letter *O*, by other as zero (Bion, 1965).

One of the most substantial contributions to psychoanalysis in terms of mathematical thinking comes from Matte Blanco, a Chilean analyst mostly known for the development of a theory of the unconscious through an application of symbolic logic. He introduced a structural distinction between the unrepressed unconscious on the one hand and consciousness and the repressed unconscious on the other, in terms of radically different types of logic according to which they operate. He linked this fundamental differentiation to the presence or absence of characteristics of the unconscious identified by Freud in relation to primary processes: absence of mutual contradiction, displacement, condensation, absence of time and substitution of psychic for external reality (Matte Blanco, 1959, pp.1–2).

Whilst consciousness and subsequently repressed material from it are ruled by asymmetric logic, the structural, unrepressed, unconscious is defined by symmetrical logic, whereby any statement can be reversed to yield its mirror opposite, with the accompanying loss of space and time that such a move entails. According to Matte Blanco, this type of unconscious process holds great complexity and is characterised by an inherent multidimensionality which makes it incompatible with consciousness and can only be unfolded into sequential thought. His observation takes into account the finding that there is no operation in three-dimensional space that would turn an object in its mirror image, but this is possible in four-dimensional space, as mathematicians August Ferdinand Möbius and Felix Klein endeavoured to explain in the 1800s (see Blacklock, 2018, for a history of this idea). As Matte Blanco explains in the introduction to his very detailed exposition published in

1975, his work on bi-logic has led him to 'considering the mind in terms of multidimensional space' (Matte Blanco, 1975, p.xx). Although he did not develop fully the link between mind and space in this book, he devoted its final part (IX) to introducing this correspondence as a promising avenue for further exploration.

As Carvalho (2010) explains, unconscious content which is not repressed is unconscious for structural reasons, as its simultaneity, complexity and multidimensional nature defy the linear, one-dimensional and sequential nature of conscious thought (p.325). The higher dimensionality of the unconscious requires sequential unfolding or translation into the fewer dimensions available to conscious thought. Something is lost and somewhat deformed in this kind of 'translation'.

Matte Blanco called the coexistence of a symmetric and an asymmetric logic a bi-logic, conceiving of the mind as functioning through the combination of at least two distinct and often polarised modes of knowing, with classificatory activity as essentially central at all levels of thought, including the unconscious (Rayner, 1995, p.2). This is why learning is both possible and difficult, given that any new notion and experience is by necessity linked to the classification system at work.

Matte Blanco developed this in a way that both resembled and departed from Bion's contribution, as he put forward two logical principles as foundation of the logic of the unconscious, namely that:

> i The thinking of the system Ucs treats an individual thing (person, object concept) as if it were a member or element of a class which contains other members; it treats the class as a subclass of a more general class, and this more general class as a subclass of a still more general class, and so on. [...]
>
> ii The system Ucs treats the converse of any relation as identical with the relation. In other words it treats relations as if they were symmetrical.
>
> (Matte Blanco, 1959, p.2)

Matte Blanco develops his intricate view of the unconscious logic in an extensive and dense volume, *The Unconscious as Infinite Sets* (1975), where he examines how the unconscious perceives in terms of infinite sets (rule *i*), as well as identifies wholes with parts (rule *ii*), through the theory of infinite sets.

Skelton (1984) questions the value of Matte Blanco's use of formal logic to develop a theory of the unconscious that can explain many diverse phenomena from a simple base, drawing attention to how, according to the theory of formal systems, 'any theory with contradictory principles explains everything and so explains very little' (p.455). In other words, the apparent explanatory power of Matte Blanco's theory is a feature that emanates from a logical error in the foundations of the model itself rather than an intrinsic quality.

Despite this rather important criticism, what remains of particular interest here is the fact that Matte Blanco's pursuit of a graphical representation of his second principle leads him to posit the relationship between a whole with more than three dimensions and its three-dimensional parts, such that in this whole more than one three-dimensional part can occupy the same space (1959, p.4). His work on this aspect of the unconscious is less known, and I explore it in more detail in the next section.

Amongst post-Freudians, Lacan remains by far the most seriously and rigorously rooted in mathematical thought, and was known to consult at length with mathematicians in formulating his later work around topology. His use of mathematics changed over time, whilst always remaining present in the formalisation of his ideas. In his early work, it was primarily with an eye for the power of mathematics as a language that he insisted on its relevance. Later on, the emphasis was in terms of structure, and yet later still in terms of an articulation of something that escapes words, but can nevertheless be circumscribed formally in mathematical terms. Thus, at various moments, he made use of set theory, game theory, functions, topology, knot theory, and was the one to introduce the use of mathemes.

According to Cléro (2002, p.45), mathematics is at the heart of Lacan's thinking, even when he does not make explicit references to it. This infusion with mathematics has not escaped criticism from both ends, with scientists questioning his grasp of the mathematics he uses and analysts questioning its relevance to psychoanalysis. While Lacan was accused of being both misguided and irrelevant in his contributions that brought together mathematics and psychoanalysis (see, e.g., Glynos and Stavrakakis (2001) for an overview on this), he managed to establish some solid foundations for further work to become possible. In its own way, this book endeavours to put forward such a contribution.

For Lacan, the key connection was in the emphasis he placed throughout his work on 'the irreducible materiality that structure entails' (Lacan, 1960, p.551), where structure is essentially symbolic in a Lacanian sense, as 'the notion of structure is by itself already a manifestation of the signifier' (Lacan, 1997, p.183), and 'the signifier is something other than meaning' (Lacan, 1997, p.184). What he designates by 'signifier' is what Freud called mnemic traces subject to retranscription (Freud, 1900).

Lacan also used mathematics to illustrate the ways in which other approaches miss the point about the unconscious as a radically different domain of experience, likening the illusion in the pursuit of ego psychology to that of the mathematician who tries to learn something about negative numbers by generating smaller and smaller positive numbers:

> The excessive prevalence of ego psychology in the new American school introduces an illusion similar to that of the mathematician [...] who having got a vague idea of the existence of negative magnitudes sets about indefinitely dividing a positive number by two in the hope of finally crossing over the zero line and entering the dreamt-of domain.
>
> (Lacan, 1993, p.166)

In other words, radical difference cannot be accessed by more (or less, in this case) of the same. As Kasner and Newman (1940) point out, existence in the mathematical sense 'is wholly different from the existence of objects in the physical world' (p.61).

Mathematically, existence is not predicated on physical location, but on the absence of self-contradiction in statements pertaining to the posited existence, such as saying that number 7 exists. In their words, 'there is no valid reason to trust in the finite any more than in the infinite' (Kasner and Newman, 1940, p.63).

Lacan's explicit statement, in his seminar of 1972–1973, on the importance of mathematics to psychoanalysis, was concise and bold: 'Mathematical formalisation is our goal, our ideal' (1988, p.119). This position does not appear to have been based on an interest in theoretical development per se, but rather it remained grounded in the clinic. Relatively early in his work, in the late 1950s, Lacan stated clearly the extent to which the use of analysis relies on a refined mathematical perception: 'a topology, in the mathematical sense of the term, appears, without which one soon realizes that it is impossible to even note the structure of a symptom in the analytic sense of the term' (Lacan, 1958, p.578). It may help, at this point, to specify that topology is 'the mathematical study of how points are connected together' (Downing, 2009, p.350). Indeed, Lacan situates the mathematical object at the intersection of the three registers of human experience that he distinguished later in his work – the Real, the Imaginary and the Symbolic[11] (Charraud, 1997, p.8). In his words, '[M]athematics [...] uses language as pure signifier, a metalanguage *par excellence*' (Lacan, 1997, p.227).

A contemporary of both Matte Blanco's and of Bion's, Lacan sought the unknown in obvious places, in plain sight, as he set off to uncover the relevance to the psyche of rules contained in something too pervasive and encompassing to be easily seen, namely, in language. He refined and emphasised the view that the unconscious is a concept built on the trail or trace left 'by that which operates to constitute the subject' (Lacan, 1964, p.703). As Paul Verhaeghe (2018) emphasises, this Lacanian unconscious is not a thing, but a concept (p.228). Rather than a compact space, or a place, with Lacan, the unconscious became understood as a locus in the mathematical sense, whereby a locus represents the set of all points satisfying some condition. Thus, in his early work, Lacan conceived of the unconscious as the discourse of the Other, an other that inhabits a kind of beyond. In cases of psychosis he located the other in 'a sort of internal beyond' (Lacan, 1997, p.123). The distinction internal vs. external is something to which Lacan later returned in terms that are explicitly both spatial and mathematical. More on this in Part III.

An interesting caution about the use of mathematics in general comes from Balibar (2003), who warns that, on occasions, mathematics has made things too easy to understand, seducing some into losing sight of conceptual inconsistencies (p.14). At the same time, mathematics has been used reliably to move away from the oversimplification of concepts through images and schemas (Binétruy, 2003, pp.66–67), a call to symbolisation and away from deceiving images, familiar to Lacanian analysts.

More recently, some analysts have addressed the question of the unknown directly, rather than through the concerns and questions raised in the clinical encounter. A notable instance is the work of Michael Eigen (2012), who explores the relationship between psychoanalysis and the unknown as he considers the Kabbalistic concept of an infinite beyond bounds and conception. This infinite God, *Ein Sof*, is fundamentally unknowable, '"buried in the depths of his negativity," i.e. refractory to attributes' (Levinas, 1999, pp.62–63). In his analysis, Eigen fondly returns time and again to a quote attributed to Sir Arthur Stanley Eddington, the physicist famed for his illuminating work on Einstein's theory of relativity: 'Something unknown is doing we don't know what' (p.18, e.g.). This mysterious space without particular orientation or markers is, however, encased in language, as Eigen illustrates with the lyrics of a childhood song: 'Hashem is here, Hashem is there, Hashem is truly everywhere; up-down, all around, that's where he can be found. Hashem literally means the Name' (Eigen, 2012, p.84).

This is the domain where something is both nothing and everything, at the same time. *Ein Sof* in Hebrew designates both the Infinite One or Infinite Nothingness (Samuel, 2007, p.90), on a level where 'distinctions and opposites vanish' (*ibid.*); in other words, beyond structure and thus beyond language. This is the energy from which, incompletely and imperfectly, ten fundamental domains (*Sephirot*) emanate, seen in Jewish mysticism as creating, nourishing and sustaining the world (Samuel, 2007, pp.288–289). Each of these ten domains is both nothing as such and infinite, and each of the ten contains all ten. The numerical allocation in the overall structure (first, second etc.) and their relative order is presented with a degree of variation across authors. For the sake of continuity, I will use here the positions and transliterations proposed by Eigen (2012).

Of the complex structure of these ten *Sephirot*, only four – the uppermost three plus a virtual one – are seen to have survived the overwhelming flow of creation (Eigen, 2012, p.88). These are Keter (the Crown), Chochmah (Wisdom) and Binah (Understanding) plus Daat (Knowledge). Knowledge only comes into existence when wisdom and understanding combine together to generate it (Samuel, 2007, p.291). Each of these domains pertains to the relationship between knowledge and the realm of all possibilities, leading to the realisation, within the overall structure, of the transmission of infinity 'into finite beings and into the finite, pluralistic realms of experience' (Samuel, 2007, p.291).

Some have treated these depictions of the interface between the divine and the ordinary experience, between possibility and realisation, as numbers, others as language. The link to numbers is through a conception of 'numerical *potentialities*, supernal and infinite in nature and origin, yet possessing the power to manifest all of creation, from the metaphysical realms to the material' (Samuel, 2007, p.292).

The four dimensions this structure encompasses are as given in Figure 2.7.

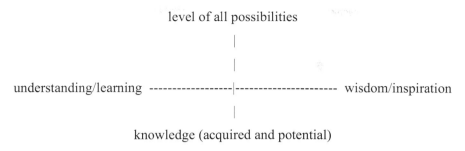

Figure 2.7 Knowledge in four dimensions.

Ordinary modes of perception and common sense are not enough to access such knowledge. Mathematics reaches beyond perception and goes further than anything towards grasping something that otherwise escapes, but even then the story remains incomplete. Like all pictorial representations, this serves to indicate the way towards an understanding that escapes representation. As Eigen (2012) emphasises, 'As with all other avenues, the tree [of Sephirot] will take you places, and if it takes you well enough along, it disappears' (p.92). The structure acts as a gateway that stops being relevant as such once it has been used.

Of all the approaches overviewed here, Eigen's one is the least formally mathematical, but it stands out in the way it links knowledge to space and dimensionality in inseparable ways.

2.2.2 The unconscious as space

Garella (2012) examines the implied spatiality of the unconscious, which 'as an object of exploration is implicitly assumed to be endowed with an extension, with some form of spatiality' (p.73). He distinguishes this from space in the usual sense, which is homogeneous and isotropic (measurements do not vary according to direction) other than at the extreme scales of the entire universe or on a quantum level. In contrast, psychic space is heterogeneous on a topographical level, with different areas of functioning having different relations to consciousness, as well as in terms of the processes that determine each space, with primary processes, defining of the unconscious, interfering with secondary ones, related to thinking (p.76). In his intricate analysis, Garella stresses the need for a specific 'translational' methodology in order to gain access to the unconscious component, although he insists that this is not in the sense of interpretation (p.71). His use of the term 'translation' is consistent with its mathematical meaning, namely, that of moving geometric objects from one location to another, without altering their properties.

Garella brings to the fore the Freudian emphasis on the surface-like nature of the psyche, as he makes references to Freud's statement in *The Ego and the Id*, namely, that the ego is 'the projection of a surface' (Freud, 1923b, p.26), explained further in an editorial note in the Standard Edition as being 'derived from bodily sensations, chiefly those springing from the surface of the body. It may thus be regarded as a mental projection of the surface of the body, besides [...] representing the superficies of the mental apparatus' (*ibid.*).

Unfortunately Garella does not seem to be familiar with Lacan's work on precisely this point, in particular the developments in seminar IX, *L'identification (1951–1952)*, where he explores how this flat nature of the psyche makes the unconscious system into a partial one (Lacan, 2020, p.132). Nevertheless, Garella thinks along similar lines, in that he recognises that the clinical practice of psychoanalysis explores psychic space, while at the same time creating this space of exploration 'whose reality is always functional or, in other words, virtual on the level of existence and therefore not ontological' (Garella, 2012, p.81). This view that the unconscious is not a complete, ready-made place is consistent with Lacan's view that the unconscious is created in analysis.

Equally, Garella seems to also miss out on another key contribution on the question of the unconscious as space, namely, that of Matte Blanco.

Space has, in Matte Blanco's view, a very important role in understanding the psyche. In his view, psychoanalysis developed within a framework resting on three key concepts: instinct (drive, in the Lacanian sense), energy and space (Matte Blanco, 1975, p.7). He goes as far as proposing that there is a natural tendency in the mind to employ the concept of space when referring to mental phenomena (Matte Blanco, 1975, p.406), yet, he argues, three-dimensional space is not sufficient to support our understanding of psychical phenomena, in the way it is possible for physical ones (Matte Blanco, 1975, p.406).

In his pursuit of the relevance of spatial thinking to psychoanalysis, Matte Blanco devotes a lot of attention to the notion of metaphor and to the bi-univocal relationship between reality and its representation, which he specifies as one of similarity, not identity (Matte Blanco, 1975, p.402). Furthermore, in his view, the very notion of structure is rooted in our concept of material space (Matte Blanco, 1975, p.407).

Matte Blanco addresses directly the relationship between space and the unconscious, taking the former as a system of relations à la Einstein, rather than as an absolute Newtonian space. He regards the body as the prime point of reference in terms of spatial orientation with regard to external physical objects, and conceives of the perception of space as inter-sensory, as he distinguishes between external space, psychological space ('the place where our experiences happen' – Rayner, 1995, p.80) and mathematical space, as defined by dimensions. While insisting that the concept of space must be taken seriously in the study of mental phenomena, Matte Blanco recognises that these 'involve spatial conceptions which are very different from those habitually employed with material phenomena' (Rayner, 1995, p.80), or as Rayner plainly translates this, 'the mind is not a bag' (p.81).

As far as mental space is concerned, Matte Blanco considers this more as a degree of freedom, in the usual sense of direction in which movement can occur. As Rayner explains, 'mental space concerns the ability of mental operations to *manoeuvre representations* of the external world, internal objects, notions and concepts' (Rayner, 1995, p.81). This indicates the space-like nature of the mind in its functioning rather than as simple geography. The ways in which mobile phones integrated into daily life and became indispensable in ways that go beyond the convenience of instant communication offer a good illustration of this. There has been irritation in the media at those who walk around engrossed in their hand-held devices, but one ought to take this state of affairs seriously. This is not an outcome created by the phones, but rather a natural inclination brought to the fore and made visible by these objects, because of what they offer: engagement with mental space to the detriment of physical space, access to a fourth dimension in preference to that limited to the three dimensions defining the space through which we move in a bodily way.

Matte Blanco explores the relationship between dimensions in terms of what happens to the representation of higher-dimensional objects in lower dimensions, namely, in terms of both the overlapping and the repetitions that this produces. Both of these have complex clinical implications. While I will explore fully the more technical aspect of this in Part II (Chapter 5, in particular), I would like to highlight here how Matte Blanco links these attributes of spatial representation to the unconscious, in particular to dreams:

> If we assume that we are following the natural human tendency (both conscious and unconscious) to apply the concept of space to thought processes, then one is struck by the fact that the contiguity and well-ordered succession of wakeful life gives way to an interpenetration of the various elements of the dream, a sort of mutually getting inside one another. In terms of three-dimensional space this appears chaotic, but if we consider the question in terms of a space of dimensions higher than three, it is no longer so. We must suppose that various dream-thoughts happen *simultaneously* in the unconscious.
> (Matte Blanco, 1975, p.417)

In other words, the time sequencing that defines our chronological, conscious mind amounts to a translation of higher spatial dimensionality. His starting point is an understanding that using the conceptual frame of space of more than three dimensions can turn apparently chaotic and incomprehensible phenomena into 'well ordained and understandable' ones (Matte Blanco, 1975, p.415). This, he argues, makes possible an understanding of the fundamental properties of the unconscious. With this in mind, Matte Blanco gives an interesting reading of Freud's second topography:

> If we consider the 'parts' of the self – id, ego, and super-ego – in terms of a three-dimensional space, but as being parts of a figure of more than three dimensions, then we may say that each of them occupies the whole of the three-dimensional volume of the person, and yet can be considered as being a part of an entity comparable to a space of more than three dimensions.
> (Matte Blanco, 1975, p.426)

In his later writing, Matte Blanco returns to the spatial description of dreams in terms as their multidimensionality as follows:

> all the spatial and temporal relations are catapulted into one another in such a way that each thing and each happening which our intellect grasps are also all other things and all other happenings.
>
> (Matte Blanco, 1988, p.265)

In other words, the whole and the parts have the particular interplay characteristic to spatial dimensions. He likens the difficulties with separating these 'parts' in three-dimensional space to a man 'trying to pour water into a jug in a painting' (p.429). Three dimensions don't go into two. At the same time, he is drawn to the relevance of space understood mathematically as a way of grasping something more about the workings of the unconscious, given that 'something which at a lower dimension is experienced as a separate object, becomes, at a higher dimension, a constituent of the whole' (Matte Blanco, 1975, p.4).

Thus, while the simultaneity and overlap characteristic of dreams cannot be represented as spatial contiguity in three-dimensional space, it can nevertheless be understood if, as he puts it, 'we assume that *the dreamer "sees" a multiple-dimensional world with eyes which are made to see only a three-dimensional world*' (Matte Blanco, 1975, p.418). What is perceived as overlap corresponds to a four-dimensional reality where two (three-dimensional) volumes can occupy the same space simultaneously (p.419).

It is also helpful to clarify here the mathematical relationship between spaces of different dimensionality which Matte Blanco himself spells out and uses in his considerations, namely, that while a space of n dimensions can 'envelop' a space of $(n + 1)$ dimensions, in the way the sides of a cube envelop the volume within, an n-dimensional space lies entirely within a space of dimensions higher than n (Matte Blanco, 1975, p.447).

Ultimately, Matte Blanco concludes that space – even in the mathematical sense – is a psychological construction, and it fits mental phenomena for that very reason, as it represents an elaboration of our sensory perceptions, stressing that 'the only objective truths we can discover are those which the structure of our nature enables us to discover' (Matte Blanco, 1975, p.325), and this nature is spatio-temporal (*ibid.*). Like Lacan, Matte Blanco is throughout careful to distinguish between identifying a phenomenon with its representation, and insists that the relationship is one of bi-univocal mapping. He concludes by inviting his readers to 'continue along this line of research and explore its full potentialities' (Matte Blanco, 1975, p.451), as he deplores the dearth of psychoanalytic thinking in spatial terms, apparently unaware of the work of his contemporaries.

A contemporary of Lacan's, Hermann explored the relationship of reciprocity between the view of space and psychic life. However, given that he sought a certain congruence between various topological spaces and psychological phenomena

(Klaniczay, 2007), it can be argued that his work remained more at the level of the unconscious *and* space, rather than the unconscious as space.

Hermann proposed that the perception we have of space is not Euclidean, but topological, at least in childhood (Hermann, 1950, p.229). He identified a three-fold link between psychic phenomena and topology, in terms of order, position and structure. The notion of 'order' relates to a representation that distinguishes between external and internal relations. This is a central tenet of psychoanalysis as seen by the IPA, to which Hermann belonged, and in which Lacan was an outlier that became excluded. 'Positions' in this context refer to the various modes of operations defining of each agency in Freud's second topography. As for 'structure', Hermann uses this to designate the congruence between space defined in qualitative rather than quantitative terms, and particular psychological phenomena. He does this by equating various types of geometry with various clinical presentations, linking Euclidean geometry to 'normality', spherical geometry to melancholy and hyperbolic geometry to schizophrenia (Darmon, 2004, p.227).

Lacan's spatial thinking is by far the most explicit and elaborate, and encompasses both the question of dimensions and that of topology. I explore in detail the idea of the unconscious in terms of spatial dimensionality in Part II, and the topological approach in Part III. Before entering that more detailed exploration, I would like to highlight briefly here Lacan's key contributions in both these respects.

Although space and topology came to occupy a central part only in Lacan's later work, he did engage with these notions from the start. As early as his seminar of 1955 he stressed 'the *miraginary* character of space' (Lacan, 1991, p.265), *miraginaire* in the original text, a hybrid notion between mirage and imaginary.

Lacan only referred to the question of spatial dimensions in passing, as he focused essentially on developing a psychoanalytic framework grounded in the foundations of topology and knot theory. Nevertheless, he went to the core of the problem when he underlined the necessity of a fourth dimension in the context of thinking about topological surfaces and knots as pathways on them (Lacan, 2020, p.341). Topologically, a path is a continuous function with a beginning and an end, and a space is defined as path-connected if any two of its points can be joined by a path (Armstrong, 1983, p.61). With these notions alone, it is possible to already see quite clearly how the world of signifiers can be thought of as a space.

Lacan's emphasis on topology is no coincidence. For him, this does not operate as a metaphor, but rather it 'shows the real of the structure' (Ragland, 2002, p.118). There is no agreement in the philosophy of mathematics as to the way in which mathematical objects are to be thought about in terms of their existence (Ferguson, 1997). Without pursuing this debate any further here, suffice it to say that topology is aligned with a structural view of mathematics, which is in turn coherent with Lacan's structural view of the unconscious, and which was the focus of his early work and was, one could argue, consolidated rather than superseded by the later, more explicitly mathematical developments in his work.

Lacan's most developed contribution to a spatial conception of the unconscious is his elaboration on the unconscious as a knot of three registers of experience, in what he calls a Borromean knot constituted by the interlinking of the Symbolic, the Real, and the Imaginary. Strictly speaking, from a mathematical point of view, this structure is not a knot, but a link[12] (i.e. a set of component loops 'tangled up together'; Adams, 2000, p.17) where the removal of any of the three rings leaves the other two unconnected to each other. In other words, each register holds the other two together. In Lacan's psychoanalytic theorising, this structure emerges as a whole, rather than being the outcome of a gradual developmental process. That is to say, this is a process of spatial transformation rather than a chronological one. With this conception of the unconscious and of the experience of being, Lacan aimed to establish not just a new concept, but a new psychoanalytic paradigm. More on this in Part III.

Bursztein (2016) argues that, through Lacan's developments of psychoanalysis in topological terms, the Freudian hypothesis of the unconscious has been reformulated as 'a purified theory of structure' (p.xviii). For Lacan, topology *is* structure (Lacan, 1973). Furthermore, Bursztein (2017) makes the bold claim that the Borromean configuration is the only way to explain the particularities of time in the unconscious, as a combination of chronology alongside points out of time. As we will see in Chapter 5, thinking of the unconscious as four-dimensional space offers a simpler way to understand the same compatibility between timelessness in the unconscious and the sequencing of chronology in lived experience.

A devoted Lacanian scholar, Verhaeghe (2001) emphasises that in Lacan's later work the subject, the unconscious, and the body appear as interchangeable (e.g. SXXII, p.50), in the sense that there is a topological homology[13] between them as spaces (Verhaeghe, 2001, p.18). The clinical implications are numerous and complex. I turn to these in Part IV.

As Blacklock (2018) stresses, 'space itself does not sit easily within any disciplinary framework' (p.5). Furthermore, no particular discipline can claim to have the ultimate formulation of space. Yet the mathematical approach remains the most rigorous, in a way that makes it possible to use its insight outside mathematics. Hence this attempt of examining its relevance to psychoanalytic thought.

The unconscious is not just structured like a language, but it is structured like a space, operating in ways defining of the interplay between spatial relations. Language itself is a kind of space, so when Lacan says that the unconscious is structured like a language, we can further understand that both are space-like structures.

Notes

1 Whilst IPA analysts use instincts and drives interchangeably, Lacanian analysts are very careful to distinguish the two, and to insist on the use of the notion of drives to designate those forces that are not aimed at satisfaction itself, but at its repetitive, circular pursuit, leaving instincts as a term which refers exclusively to biological needs (e.g. Evans, 1996).

2   In the same year, at the November meeting of the Vienna Psychoanalytic Society, Freud is recorded as conceptualising the unconscious in spatial terms, referring to 'the peculiar phenomenon of opposite trends existing side by side, which has in a sense the quality of space' (Nunberg and Federn, 1974, p.308).
3   Wahrnehmungen (perceptions).
4   Wahrnehmungszeichen (indication of perception).
5   Unbewusstsein.
6   Vorbewusstsein (preconsciousness).
7   Bewusstsein.
8   According to footnote 3 on page 72 of SE22, this abbreviation is only used one more time by Freud, 16 years later.
9   Freud used both 'instinct' and 'drive', yet the translation of the Standard Edition does not retain this important distinction, using 'instinct' to mean both. In his return to Freud, Lacan reinstated this differentiation between the fixity of biological needs carried by instinct and the essence of drives as highly subjective, and therefore variable, seeing as they are shaped by the contingency encountered by each individual in their life history.
10 As Kasner and Newman (1940) point out, the word 'function' has a multitude of meanings, and the specific mathematical sense cannot be guessed from the others. They explain its meaning as that of a table that gives the relation between sets of variables where the values of one variable are determined by the values of the other, and illustrate it thus: 'one variable may express in decibels the amount of noise made by a political speaker, and the other, the blood pressure units of his listeners' (p.5).
11 These central Lacanian concepts will appear throughout. While familiar to some, the words themselves are potentially misleading to those new to Lacanian literature, as each of them is used here in a sense that is both specific and substantially different from common use. Also, as is often the case in analytic literature, the way in which these terms are used has evolved over time, both in Lacan's work and in that of those who took it further. In brief, Lacan uses these terms to designate three types of order, or three registers, according to which all psychoanalytic phenomena and human experience could be described. In brief, the Imaginary is most closely linked to animal psychology, and pertains to the domain of image and its illusion; the Symbolic is the realm of language and structure according to differentiated, discrete elements known as signifiers, whereas the Real operates as some kind of remainder, undifferentiated and without fissure (Lacan, 1991, p.97), and fundamentally unknowable (Evans, 1996, pp.159–161). Given that I also use the mathematical concepts of real and imaginary (in relation to number theory), all uses in the Lacanian sense will be capitalised, to ease differentiation.
12 A knot is regarded as a link of one component (Adams, 2000, p.17), whereas the Borromean structure he uses consists of three interconnected unknots.
13 Strictly speaking, this is not topologically accurate. Might he have been referring to a homeomorphism, instead? Homology denotes a rigorous mathematical method for defining and categorising holes in a manifold, i.e. topological spaces that locally resemble Euclidean space.

References

Adams, D. (1979). *The Hitchhiker's guide to the galaxy*. London: Arthur Barker.
Akhtar, S. and O'Neil, M.K. (eds.) (2013). *On Freud's "The Unconscious"*. London: Karnac.
Armstrong, M.A. (1983). *Basic topology*. New York: Springer-Verlag.
Balibar, F. (2003). Le réel a toujours eu quatre dimensions. In Cohen-Tannoudji, G. and Noël, È. (eds.). *Le réel et ses dimensions*. Editeur: EDP Sciences, pp.11–23.

Bateman, A. and Holmes, J. (1995) *Introduction to psychoanalysis: Contemporary theory and practice*. London: Routledge [reprinted in 2008].
Binétruy, P. (2003). Les nouvelles dimensions de l'Univers. In Cohen-Tannoudji, G. and Noël, È. (eds.). *Le réel et ses dimensions*. Editeur: EDP Sciences, pp.57–67.
Bion, W.R. (1962). *Learning from experience*. London: William Hienemann Medical Books Ltd., reprinted by Karnac, 2007.
Bion, W.R. (1963). *Elements of psycho-analysis*. London: William Hienemann Medical Books Ltd., reprinted by Maresfield Reprints, 1984.
Bion, W. R. (1965). *Transformations*. London: William Heinemann [Reprinted London: Karnac Books 1984].
Blacklock, M. (2018). *The emergence of the fourth dimension: Higher spatial thinking in the fin de siècle*. Oxford: Oxford University Press.
Braithwaite, R.B. (1953). *Scientific explanation: A study of the function of theory, probability and law in science*. Cambridge: Cambridge University Press.
Bria, P. (1981). Catastrophe and transformations. *Rivista di Psicoanalisi*, 27, pp.503–512.
Bursztein, J.-G. (2016). *On the difference between psychoanalysis and psychotherapy*. Paris: Nouvelles Etudes Freudiennes.
Bursztein, J.-G. (2017). *L'Inconscient, son espace-temps: Aristote, Lacan, Poincaré*. Paris: Hermann.
Carvalho, R. (2010). Matte Blanco and the multidimensional realm of the unconscious. *British Journal of Psychotherapy*, 26(3), pp.324–334.
Charraud, N. (1997). *Lacan et les mathématiques*. Paris: Anthropps.
Cléro, J-P. (2002). *Le vocabulaire de Lacan*. Paris: Ellipses.
Darmon, M. (2004). *Essais sur la topologie lacanienne*. Paris: Association Lacanienne Internationale.
Downing, D. (2009). *Dictionary of mathematics terms*, 3rd ed. New York: Barron's Educational Series.
Du Sautoy, M. (2016). *What we cannot know*. London: 4th Estate.
Eigen, M. (2012). *Kabbalah and psychoanalysis*. London: Karnac.
Etchegoyen, H.R. (1999). *Fundamentals of psychoanalytic technique*. London: Karnac.
Evans, D. (1996). *An introductory dictionary of Lacanian psychoanalysis*. London: Routledge.
Ferguson, S. (1997) What is the philosophy of mathematics? *Philosophy Now*, 19.
Freud, S. (1895). Studies in hysteria. SE1.
Freud, S. (1896). Letter 52 from extracts from the Fliess Papers. The standard edition of the complete psychological works of Sigmund Freud, Volume I (1886–1899): Pre-psychoanalytic publications and unpublished drafts, pp.233–239.
Freud, S. (1900). The interpretation of dreams. SE4 & SE5.
Freud, S. (1905a). Jokes and their relation to the unconscious. SE8.
Freud, S. (1905b). Three essays on the theory of sexuality. SE, 7, pp.123–243.
Freud, S. (1911) *Formulations on the two principles of mental functioning*. In: Standard Edition. Vol.XII. London: Hogarth Press, 1958. pp.213–226.
Freud, S. (1912). A note on the unconscious in psychoanalysis. SE12, pp.257–266.
Freud, S. (1915a). Repression. SE14, pp.141–158.
Freud, S. (1915b). The unconscious. SE14, pp.161–215.
Freud, S. (1920). Beyond the pleasure principle. SE18, pp.2–64.
Freud, S. (1923a). Two encyclopedia articles. SE18, pp.235–259.

Freud, S. (1923b). The ego and the id. SE19, pp.3–66.
Freud, S. (1933). New introductory lectures on psychoanalysis. SE22, 57–111.
Freud, S. (1937). Analysis terminable and interminable. SE23, 211–253.
Garella, A. (2012). Exploration of the unconscious: Some considerations on space, the object and the process of knowledge in psychoanalysis. *The Italian Psychoanalytic Annual*, 6, pp.73–89.
Glynos, J. and Stavrakakis, Y. (2001). Postures and impostures: On Lacan's style and use of mathematical science. *American Imago*, 58, pp.685–706.
Grose, A. (2014). The unconscious from Freud to Lacan. In Gessert, A. (ed.) *Introductory lectures on Lacan*. London: Karnac, pp.1–21.
Hermann, I. (1950). Rapports spatiaux de quelques phenomenes psychiques. *Acta Psychologica*, 7, pp.225–246.
Kasner, E. and Newman, J. (1940). *Mathematics and the imagination*. New York: Dover Publications.
Klaniczay, S. (2007). Espace et psyché: À la mémoire de Imre Hermann. *Le Coq-héron*, 188, pp.35–41.
Lacan, J. (1953). The function and field of speech and language in psychoanalysis. In Fink, B. (ed.) (2002). *Écrits*. London: WW Norton, pp.541–574.
Lacan, J. (1958). The significance of the phallus. In Fink, B. (ed.) (2002). *Écrits*. London: WW Norton, pp.575–584.
Lacan, J. (1960). Remarks on Daniel Lagache's presentation. In Fink, B. (ed.) (2002). *Écrits*. London: WW Norton, pp.541–574.
Lacan, J. (1964). Position of the unconscious. In Fink, B. (ed.) (2002). *Écrits*. London: WW Norton, pp.703–721.
Lacan, J. (1973). L'Étourdit. *Scilicet*, 4, pp.5–52.
Lacan J. (1988[1975]). *The seminar. Book XX. Encore: On feminine sexuality, the limits of love and knowledge, 1972–1973*. London: WW Norton.
Lacan, J. (1991[1978]). *The seminar. Book II. The ego in Freud's theory and in the technique of psychoanalysis, 1954–1955*. London: WW Norton.
Lacan, J. (1997 [1981]). *The seminar. Book III. The psychoses, 1955–1956*. London: WW Norton.
Lacan, J. (2020). *L'identification: Séminaire 1961–1962*. Éditions de l'Association Lacanienne Internationale. Paris: Publication hors commerce.
Langs, R. (1988). Mathematics for psychoanalysis. *British Journal of Psychotherapy*, 5, pp.204–212.
Laplanche, J. and Pontalis, J.B. (1988). *The language of psychoanalysis*. London: Karnac Books.
Levinas, E. (1999). *Alterity and transcendence*. London: Athlone Press.
Matte Blanco, I. (1959). Expression in symbolic logic of the characteristics of the system Ucs or the logic of the system Ucs. *International Journal of Psychoanalysis*, 40, pp.1–5.
Matte Blanco, I. (1975). *The unconscious as infinite sets*. Aylesbury: Duckworth.
Matte Blanco, I. (1988). *Thinking, feeling, and being: Clinical reflections on the fundamental antinomy of human beings and world*. London: Routledge.
New Scientist (2016). https://www.newscientist.com/article/2083397-the-play-x-will-have-you-clock-watching-but-in-a-good-way/ Accessed 8 April 2016.
Nunberg, H. and Federn, E. (eds.) (1974). *Minutes of the Vienna psychoanalytic society 1910–1911*, vol.3. New York: International Universities Press.

Ragland, E. (2002). The topological dimension of Lacanian optics. *Analysis*, 11, pp.115–126.
Rayner, E. (1995). *Unconscious logic: An introduction to Matte Blanco's bi-logic and its uses*. London: Routledge.
Rose, J. and Shulman, G. (2016). *The non-linear mind: psychoanalysis of complexity in psychic life*. London: Karnac.
Rycroft, C. (1972). *A critical dictionary of psychoanalysis*. London: Penguin Books.
Samuel, G. (2007). *The Kabbalah handbook: A concise encyclopedia of terms and concepts in Jewish mysticism*. New York: Jeremy P. Tarcher/Penguin.
Sandler, J. et al. (1997). *Freud's models of the mind: An introduction*. London: Karnac [reprinted in 2005].
Skelton, R. (1984). Understanding Matte Blanco. *International Journal of Psychoanalysis*, 65, pp.453–453.
Soler. C. (2014). *The unconscious reinvented*. London: Karnac.
Teissier, B. (1997). Des modèles de la Morphogénèse à la morphogénèse des modèles. Republié dans *"La Cause Freudienne"*, 37.
Verhaeghe, P. (2001). Subject and body: Lacan's struggle with the Real. In Verhaeghe, P. (ed.). *Beyond gender: From subject to drive*. New York: Other Press, pp.65–97.
Verhaeghe, P. (2018). Position of the unconscious. In Vanheule, S., Hook, D. and Neil, C. (eds.). *Reading Lacan's Écrits: From 'Signification of the phallus' to 'Metaphor of the subject'*. London: Routledge, pp.224–258.
Wells, D. (1991). *The Penguin dictionary of curious and interesting geometry*. London: Penguin Books.
Wollheim, R. (1991). *Freud*. 2nd ed. Glasgow: Fontana Press.

Part II

The unconscious as inaccessible space

Both Freud and Lacan emphasised the discontinuity and the lack of symmetry between conscious and unconscious processes and aspects of being. In Freudian terms, the unconscious is not a kind of consciousness of which we are not aware yet, it is not 'sub-conscious', it is not just a 'yet unknown', but a kind of unknown that is not directly knowable. The processes that dominate this realm are radically different from those of the conscious operation of the mind. This distinction was most clearly elaborated by Freud in his early work, namely in the *Project for a scientific psychology* of 1895, and in the *Interpretation of dreams* of 1900, although, as Laplanche and Pontalis (1988) stress, it remained 'an unchanging coordinate of his thought' (p.339).

Freud described 'two fundamentally different kinds of psychical process' (Freud, 1900, p.597), and attributed them to distinct systems of the psychical apparatus, which he called Unconscious (*Ucs*) and Preconscious (*Pcs*). He did not give a special place to consciousness, merely pointing out that 'the most complicated achievements of thought are possible without the assistance of consciousness' (Freud, 1900, p.593). Instead, he concentrated on the interplay between the unconscious and preconscious systems, each governed by a distinct set of rules: the un/pleasure principle and the reality principle, respectively. Thus, the unconscious system has the aim of avoiding unpleasure and of obtaining pleasure, both of which are defined in terms of quantities of excitation: too much tension produces unpleasure, and pleasure follows from its release or discharge, or from its avoidance. The pleasure principle (to which Freud initially referred as the unpleasure principle) is coupled and contrasted with the reality principle, which comes into play only gradually, as the unconscious fails to deliver satisfaction purely through hallucination, and other routes to attaining it, in accordance with the conditions imposed from the external world, become both necessary and possible.

As Laplanche and Pontalis emphasise, '[V]iewed from the economic standpoint, the reality principle corresponds to a transformation of free energy into bound energy' (1988, p.379). In Freud's language, the energy in the system *Ucs* is free or mobile, while the one in *Pcs* is bound or quiescent. If we listen to this in spatial terms, then *Pcs* has at least one less dimension compared to the *Ucs*, or fewer degrees of freedom.[1] Movement that is possible in one, is not available in the other. Whatever enters into consciousness, does so via *Pcs*, in a way that entails a necessary loss, as does any move from a space of a higher dimensionality to a spaces with fewer dimensions. Something gets squeezed and flattened in this process of transcription. If we take the example of dreams alone, such as we recall them, they are often contradictory, disjointed, nonsensical or weird, as measured against standard thought and perception in waking life, although they retain an internal consistency which appears to be governed by another kind of logic. The fluidity of movement that is possible in the *Ucs* arrives as stilted and disjointed to our consciousness. This is consistent with the fact that, spatially, something that can appear as fragmented and heterogeneous in a lower dimension, can constitute a coherent whole in a higher dimension. Thus, the discrepancy between the two systems introduced by Freud can be understood in terms of differences in dimensionality between two spaces. The inaccessibility of the unconscious from the direction of the preconscious is consistent with the problems encountered spatially with movement from a space with lower to one with higher dimensions, a fundamental asymmetry we are going to return to in some detail in Chapter 4.

The mechanisms at play in unconscious operations, or in primary processes, are those of displacement, condensation and overdetermination. **Displacement**, in Freudian terms, relates to the movement of free or unbound energy in the *Ucs* system, and results from its capacity to travel along pathways defined as chains of association (Laplanche and Pontalis, 1988, p.121). Displacement is, in other words, a spatial quality related to movement. **Condensation** is an operation that brings together a multitude of energies, creating a confluence of various pathways, be it by the recurrence of one element in an unconscious process, by the creation of a composite figure, or by overlapping of images that blurs some outlines, while stressing others (Laplanche and Pontalis, 1988, p.83). In spatial terms, condensation is akin to projections from higher onto lower dimensions, characterised by repetition and overlapping. By the nature of its operation, the mechanism of condensation is closely linked to overdetermination. As Laplanche and Pontalis (1988) clarify, there are two ways of understanding ***overdetermination***: (a) as the multiplicity of causes behind an unconscious formation; (b) as a multiplicity of unconscious elements which 'may be organised in different meaningful sequences' (p.292). While the first version is purely causal, with causality usually seen as operating along the temporal axis, where the cause precedes the effect, the second one can be understood in spatial terms, whereby the elements in question can occupy a variety of positions relative to each-other, producing multiple possible landscapes. The meaning involved relates to either the conscious order that 'makes sense' of such elements, or to the dream-like logic that unifies them in ways that are different from those offered by consciousness.

A rigorous follower of Freud, Lacan not only emphasised the radical alterity of the unconscious, but also expressed this in spatial terms. In his 1966 paper on the

Position of the Unconscious (the title itself loaded with spatial connotations), he specified '*the place* from which it could speak' (p.710), as

> the entrance to the cave, towards the exit of which Plato guides us, while one imagines seeing the psychoanalyst entering there. But things are not that easy, as it is an entrance one can only reach just as it closes [...], and the only way for it to open up a bit is by calling from the inside.
>
> (p.711)

'*It*' is the unconscious, and this spatial access is impossible as such, occurring at best in the form of points of contact rather than of full movement (another degrees of freedom problem), it is intermittent and fragmented, a kind of pulsation or 'beat' which is 'inscribed in a geometry in which space is reduced to a combinatory: it is what is called an "edge" [*sic!*] in topology' (p.711), a kind of space that one cannot enter, but merely demarcate. Lacan stresses here that the nature of this space is such that it is improper to try to turn it into an inside (*ibid*).

The original word Lacan used was '*bord*', and the corresponding topological concept is that of boundary, and not of edge, as the Bruce Fink translation has it. This distinction is important, because the concept of edge also exist, but in Euclidean geometry rather than in topology, so defining the demarcation of a radically different type of space from the one to which Lacan refers here. Edge is the term most commonly used to designate a line segment that connects two vertices in Euclidean geometry (e.g. the side of a triangle), where it can also be seen as a boundary. But in more general terms, mathematically, a boundary is the set of points which can be approached both from within a space and from the outside of it. Boundary points belong to the space, but are not inside it. In this sense, Lacan regards the unconscious as a (mathematical, topological) boundary, a linear locus, a kind of space without an interior. As Vappereau (2006) rightly emphasises, for an open space, the boundary is extrinsic, but simply because it is not intrinsic, it does not mean that it does not exist. Closed spaces,[2] as we will see in Part III, do not have a boundary.

In other words, the unconscious is not a space that we can enter, not somewhere we can go, but a structure that is organised spatially and about which we can posit that it has a certain continuity and compactness, in the way that Lacan did when referring to the Real. In anticipation of the more technical treatment of these notions in later chapters, suffice it to say now that compactness refers to 'a sort of completed infinity' (Barr, 1964, p.95), a property that a sphere has, unlike a flat plane, even if they both have the same infinite number of points.[3] The unconscious is not an accessible space. Mathematically, boundary spaces[4] do exist, and Lacan posited the unconscious as one such configuration. But a boundary is also, potentially, a place of intersection, a commonality, between a higher-dimensional space and a lower-dimensional space. It is this possibility that I am particularly interested in exploring here.

The movement between consciousness and the unconscious is not akin to a movement from one location to another in a kind of continuous space defined by three-dimensional Euclidean geometry, the most familiar model of space in the every-day sense. Also, because of the radical difference in the nature of these psychic agencies understood as spaces, one cannot expect the possibility of a one-to-one correspondence between the two: one cannot be mapped onto the other. It is not like walking from one room to another, so to say. We know this through our daily experience – either as patients or as clinicians. What is of interest here is how a mathematical understanding of the kind of space that might be involved can help us think about the unconscious in new ways, both theoretically and clinically.

In terms of mathematics, space can be thought about in two fundamental ways: as rigid, smooth space, and as stretchy, crumpled space. Smooth space is the space of even, flat, straight lines and tidy curves, as studied by Euclidean geometry since around 300BC. This is the space of Newtonian physics, and of Freudian explorations. The appeal of this formulation about space is linked to the limitations to perception imposed by our senses, in particular by our eyes: 'Euclid's geometry is founded on lines and points, which are among the basic structures that the visual cortex extracts by processing incoming sensory data.' (Stewart, 2017, p.35). Nevertheless, '[E]mpirical observation has shown that real space isn't Euclidean' (Stewart, 2017, p.46). Crumpled space, is a more recent conception, associated with the development of topology (rubber[5]-sheet geometry) since the 1700s.[6] This is the space of Einstein's physics, and of Lacanian developments in psychoanalysis.

Different conceptions of space allow for different possibilities of movement. In the next two chapters I will attend to the Euclidean understanding of space, with particular emphasis on dimensionality, turning to topological spaces and the notion of impossibility in Part III.

Remaining within the realm of Euclidean space for now, the notion of inaccessible space can be understood in two ways: either in relation to infinity, or in relation to dimensions.

Inaccessibility expressed in terms of infinity is consistent with the notion that space is homogeneous and extends infinitely far in all possible directions. In such a space, something always remains out of reach because it is further away. At the same time, the underlying view – based on mistaking infinity for "very large" - is that, whilst things might be very far, we could reach them eventually, at least in principle. Psychoanalytically, this is consistent with Freud's earlier conception of suffering linked to repressed material which could be retrieved a bit like an archaeological artefact, if one were to dig deeply enough, carefully enough, for long enough.

Yet Freud was never far from the recognition that it was not as simple as that, that a kind of "beyond" always remained, something fundamentally inaccessible. If that is the case, the work of analysis is not a matter of continuing until something could be exhausted, at least in principle, but rather it is an instance of dealing with something that cannot ever be accessed through more of the same. Also, clinically, change at the level of the unconscious does not simply occur by doing more of the same, but rather requires for something to be shaken up, or disrupted, in a more radical way, or cut, as Lacan carefully set out throughout his work.

Freud's specific formulation of this point has a spatial connotation, albeit still in the 'vertical' sense, with the repressed as deep below:

'There is often a passage in even the most thoroughly interpreted dream which has to be left obscure; this is because we become aware during the work of interpretation that at that point there is a tangle of dream-thoughts which cannot be unravelled and which moreover adds nothing to our knowledge of the content of the dream. This is the dream's navel, the spot where it reaches down into the unknown.

(Freud, 1900, p.525)

This particular type of inaccessibility can be best expressed mathematically in terms of dimensions. Movements from a higher to a lower dimension are mathematically known as projections. A three-dimensional object lit from behind casts a two-dimensional shadow on a flat surface. This movement from three to two dimensions comes at a price of loss of information. What is differentiated in the third dimension, appears as homogeneous and indistinct in the shadow which occupies a space which lacks one lost dimension. Hence the 'tangle' in Freud's description. The way we can conceptualise this in spatial terms is by positing that what matters is located in another dimension.

His reference to the "tangle of dream-thoughts" also resonates with knot theory, some of which was taken seriously and examined by Lacan several decades later. I will say more about this in Part III. For now, it is worth noting that a knot that cannot be undone in three-dimensional space can be easily undone when one extra dimension is added.

One thing that all analysts agree upon, irrespective of their orientation, is that something about the unconscious remains fundamentally out of reach, which implicitly gives it a spatial quality. Seen by some as a hidden place and by others as a function that operates in a kind of beyond, the unconscious cannot be apprehended directly; instead, it is knowable through its effects, which are most visible in every moment, in particular at times of suffering.

To recap, mathematically, something out of reach can be formalised as either some position to which cannot be attained fully (infinity), or as a location that belongs to a space which is more complex than the one in which one operates (dimensionality). In the next two chapters. I will explore each version of spatial inaccessibility in turn, inasmuch as they can illuminate one possible spatial conception of the unconscious as inaccessible. Specifically, Chapter 3 introduces the notions of infinity and incompleteness through an exploration of key ideas from number theory, linking these to the Freudian formulation of the unconscious. Chapter 4 takes further the ideas of dimensionality encountered in the world of numbers, and presents a possible reading of the unconscious as space. Chapter 5 expands this approach further, to investigate what remains as inaccessible as the relation between spaces of unequal dimensions.

Notes

1 The number of dimensions of a particular space is identical with the number of degrees of freedom of a point in that space (Matte Blanco, 1988, p.294).
2 A topological space consists of a set of elements together with a particular structure (called a topology). A closed set is one which contains its own boundary.
3 In topology, this property translates into the possibility of partitioning such a space into a finite number of triangles, or triangulation (Adams, 2000, p.83).
4 In most general terms, on a surface, the boundary of a piece of a surface is the curve which separates that piece from the rest of the surface (Arnold, 2011, p.63).
5 Such a rubber deformation is called an isotopy and it can be applied to either surfaces or knots. Two surfaces or knots are isotopic if they are equivalent under a rubber deformation (Adams, 2000, p.73).
6 According to Armstrong (1983), the idea of explaining topological equivalence by thinking of spaces made of rubber is attributed to Möbius (p.11).

References

Barr, S. (1964). *Experiments in topology*. New York: Dover Publications.
Freud, S. (1900). *The interpretation of dreams*. SE4 & SE5
Laplanche, J. and Pontalis, J. B. (1988). *The language of psychoanalysis*. London: Karnac Books.
Stewart, I. (2017). *Infinity: A very short introduction*. Oxford: Oxford University Press.
Vappereau, J. M. (2006). La D.I. http://jeanmichel.vappereau.free.fr/textes/La%20DI.pdf

Chapter 3

The unconscious as infinity and the possibility of incompleteness

With reference to the infinite quality of space, Newton wrote: 'You may want to object that we can't *imagine* that there is infinite extension. I agree! But I contend that we can *understand* it' (Newton, 1666, p.9). The Freudian unconscious is such a space, one which we cannot perhaps imagine, but one we can, nonetheless, understand something about. One attribute of infinity is that it cannot be exhausted. As a consequence, the work of analysis is without an end in the usual sense, even though each clinical encounter is ultimately finite and each analysis has at least a beginning.

What I want to consider here is conceptions of space that rely on the notion of infinity and examine how they impact on the view that something can be known, if not actually, at least in principle. In this sense, the possibility of completeness is maintained, even if only as potentiality. Nevertheless, the character of such spaces raises the question of gaps and discontinuities, to which I turn in Part III. Throughout this exploration, it is important to keep in mind that, mathematically, infinity is not another way of expressing the notion of something very large but, as Stewart (2017) stresses, 'the absence of any limit' (p.19). However large a number, it is not ever infinite. The two notions are radically different, in the same way that consciousness and the unconscious are. I will not consider here philosophical and theological notions of infinity (see, e.g., Levinas, 1999, for a compact and relevant overview, in particular the chapter titled *Infinity*), but only emphasise that this is a meta-concept encompassing a variety of ideas, and that for the purpose of this analysis, this definition, of the absence of limits, suffices.

The essence of infinity rests in a particular property, namely, that the whole is no greater than some of its parts (Kasner and Newman, 1940, p.43), and that 'an infinite class may contain as proper parts, subclasses equivalent to it' (p.56). Formally, '[A]n infinite class is one which can be put into one-to-one reciprocal correspondence with a proper subset of itself' (p.57). This is essentially consistent with the analytic experience, where no matter where we start, the full structure of the subjective experience unfolds without the need to speak of everything. Every session and every encounter maps onto the totality of the subjective experience, and vice versa. At the same time, every encounter and articulation has at its core an incompleteness, as does the lived experience of the subject.

42 The unconscious as inaccessible space

I begin here the exploration of possible rigorous approaches to a structure where something is intrinsically both included and out of reach, in the way that daily experience shows us that the unconscious is positioned in relation to the experience of being. The starting point is the world of numbers, a space whose structure illuminates the relevance and limitations of thinking about the unconscious as a limit that can be gradually approached but never reached as such. This opens up a systematic understanding of the interplay between what can be symbolised, what has not yet been represented and what escapes symbolisation altogether, leading into the next step of the enquiry, in terms of dimensionality, in the following chapter.

3.1 Counting, recurrence and the spaces in between

I commence by considering a particular kind of space: an infinitely long straight line. This is a one-dimensional space, the lowest dimensionality in which some movement is possible. The line I have in mind is not any line, but the one known as the number line, the familiar abstract representation of real numbers that one encounters in basic mathematics.

The world of numbers is the simplest formalised space which can be thought about either in metric terms, with focus on precision and measurement, or in a topological sense, with attention to the relative positions of the elements it comprises. The questions of what is located where and of whether there is something everywhere in this space are central to conceiving of its spatial structure. A closer examination shows that what can be located is the minority of elements, and that in the gaps in between these clear markers, unfathomable depths can open, sometimes hinting at discontinuity and therefore at impossibility. It is this second spatial sense, as a topology, that makes the world of numbers relevant to the unconscious. As signifiers, numbers show that what can be inscribed is always partial and fails to capture what remains outside language, whilst being encountered at the level of experience.

The number line is akin to a ruler that extends infinitely in both directions, with every point assumed to correspond to a real number, and every real number assumed to correspond to a point on this line. Real numbers are real as opposed to imaginary – not in the Lacanian, but in the mathematical sense. They constitute the set of all numbers with a finite or infinite decimal expansion (Gowers, 2008, p.18). Real numbers were initially regarded as possible numbers, while imaginary ones as impossible, the term 'imaginary' having been used for a long time in a derogatory sense. In terms of their function, imaginary numbers were akin to a construction in analysis, or to the positing of the existence of the unconscious itself: once considered, new ways of understanding and new possibilities of transformation emerged. More on imaginary and complex numbers later.

Usually depicted with arrow ends pointing in opposite directions, and centred on zero, the image of the number line is the image of a finite segment that points at infinity (Figure 3.1).

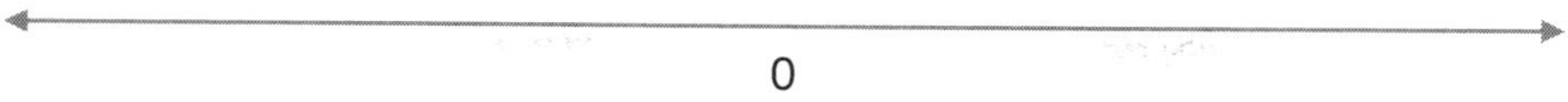

The unconscious as infinity 43

$$\longleftarrow \underset{0}{\rule{8cm}{0.4pt}} \longrightarrow$$

Figure 3.1 The number line.

In The Book of Sand, Borges (1998[1975]) tells the story of a man who is unexpectedly forced to find a place for infinity itself. The story begins thus: 'The line consists of an infinite number of points; the plane, of an infinite number of lines; the hypervolume, of an infinite number of volumes…' (p.89). This is a story about dimensions. The nameless man of this story lives on his own and has a large collection of Bibles. One day he parts with one of these in exchange for another sacred book, titled Holy Writ. The book he thus acquires is hefty, the pages printed with unfamiliar characters, and numbered in no discernible sequence. The mysterious seller tells him: 'Look at it well. You will never see it again' (p.90). Indeed, this is the Book of Sand which has neither beginning nor end, the number of its pages is 'literally infinite', so that they can be numbered 'any way whatever' (p.91). The buyer spends time trying and failing to retrieve illustrations glimpsed once, works hard to decipher the structure that rules the makings of the book. Unable to do so, he begins to see it as 'a nightmare thing, an obscene thing, and that it defiled and corrupted reality' (p.93). Afraid that burning it might produce infinite smoke that would 'suffocate the planet', he decides to hide it in a library, amongst maps and periodicals. The best place to hide infinity is in the spaces between finite things, between inscriptions of space and markers of time. This is precisely what the structure of the number line reveals, namely, the possibility of a space where discernible entities delineate subspaces of infinite depth and complexity.

In another work of fiction, just as seriously rooted in mathematics, Stewart (2001) sums up this view of infinity in the description of the sign welcoming visitors to Infinityville: 'Where Things Happen that Don't' (p.113).

As we are going to see, most of the numbers world is located in the spaces between the numbers we know something about, in the gaps between what can be written. Because, although the line number can be written, and numbers can be marked on it as if on a ruler, what is of interest here is not what can be represented on the line, but the incompleteness its makeup sustains and evokes. This is very much of the same nature with the counting that Lacan refers to with regard to repetition: what is marked, what can be marked, is not that which is pursued (Lacan, 2020). The number line hints at something missing, at a gap in the form of incompleteness. What matters even more than what we can mark on this ruler-like line is what happens in the spaces between the marked points, that is to say, that which escapes representation. In his exploration of infinity in relation to psychoanalysis, Eigen (2012) illustrates this particular interplay with reference to Kafka, who 'called his whole life an incomplete moment' (p.71).

Given the nature of the space under consideration, two main questions arise. First, is there a number at every point on the line? Second, are there numbers without a corresponding point on the line?

Indeed, the number line and the question of continuity it poses was of great interest to generations of mathematicians, and it links to the distinction between discrete and continuous numbers. Integers are discrete numbers, and there are gaps between them on the number line, as there is no natural number between 1 and 2, between 2 and 3, etc. The same cannot be said for real numbers: between any two real numbers there are infinitely many other real numbers (Stewart, 2017, pp.28–29), the points on a line are 'everywhere dense' (Kasner and Newman, 1940, p.54). The set of real numbers, as a topological space, is complete, by construction: it includes rational numbers and (irrational) ones that represent all the potential limits of all sequences of rational numbers (Vivier, 2004, p.112). Rational numbers are numbers that can be expressed as a ratio between two other numbers, but most numbers cannot be expressed in this form. The latter are numbers that densely occupy the space between rational numbers, and are known as irrational numbers, not because they are lacking reason, but because they cannot be obtained as the result of a ratio between two integer numbers. Given that irrational numbers are significantly more numerous than rational ones, it follows that the possible limits or points of convergence[1] of all such sequences outnumber the possible elements of the sequences. In other words, in the world of numbers, there are substantially more terminal points than intermediate steps. The question of mapping of numbers onto a straight line 'infinitely rich in point-individuals' was carefully examined by Richard Dedekind, a contemporary of Freud's, who came to define each number as a unique cut in the continuity of this infinity (partial translation published in Smith, 1959, pp.35–45). This resonates with the definition of dimensions as cuts given by Poincaré (1912), and also with Lacan's considerations around the subject as something residing the gaps in language. More on this in Chapter 4.

Amongst the mathematicians devoted to this domain, Georg Cantor – also a contemporary of Freud's – and his contributions attracted the attention of psychoanalysts interested in the question of the relation between the unconscious and infinity, and in the place that discontinuity and incompleteness have in our lives (see Charraud, 1994).

Cantor was particularly preoccupied with the question of continuity in the space we are considering here, the line of real numbers. What he was keen on examining was the topology of a line formed of a set of points, on understanding the relationship between the points and the line. This was, in fact, a question about the relationship between dimensions – as by definition points are zero-dimensional, and lines unidimensional – a question he never quite managed to resolve in a way that reassured his own intense unease about any possibility of discontinuity. Instead, it was easier for him to examine the question of infinity, of endlessness, as opposed to that of gaps, as he considered something about the 'size' of the class of real numbers and therefore that of points on a line segment (Kasner and Newman, 1940, p.55). Cantor was the one to introduce the notion of transfinite numbers that went beyond infinity as understood until that moment. One such number is the cardinality of the set of real numbers, which is a transfinite number, a number larger than the infinite number of natural numbers (a kind of measure of size, of its numerosity), that is to say, of positive integers: 1, 2, 3... A meta-number, one could say. The set of integers could not have its magnitude captured in a number of the kind it includes; the measure had to come from outside. In this sense, for Cantor, infinity had a beyond, a space that could at least be named, if not exactly located or measured. In this regard, Kasner and Newman (1940) formulate an important caution:

> There is no point where the very big starts to merge into the infinite. You may write a number as big as you please; it will be no nearer the infinite than the number 1 or the number 7. Make sure that you keep this distinction very clear and you will have mastered many of the subtleties of the transfinite.
>
> (p.34)

The first childhood experience of numbers is that of counting. It is perhaps no coincidence that, historically, counting was the main context in which the notion of infinity emerged, in relation to the idea of movement towards larger and larger numbers (Stewart, 2017, p.23). Indeed, recurrence is a principle for building infinite series, a way of structuring infinity: from any integer number n, it is possible to construct the next integer by adding 1, to arrive at $n + 1$. In this respect, '[T]he mathematics of the infinite is a sheer affirmation of the inherent power of reasoning by recurrence' (Kasner and Newman, 1940, p.36).

Numbers are a way of both capturing and limiting an experience though measurement, as well as a way of hinting at what escapes such measurement. The most common of all numbers are integers, the so-called natural numbers, 1, 2, 3 ...., where each number is safely and predictably followed by a number greater by 1. Zero is not included in the set of natural numbers, but it is part of the set of whole numbers. Even if we would run out of time, the idea is that we could conceivably 'count to infinity', and at any point we would know what comes next. If we added up all these increasingly large numbers, most would argue that the result would be infinity itself. There is something appealing to the thought that something we cannot grasp the size of (space, time) can go on forever (Stewart, 2017). Yet a particularly surprising caution about this reassuring view came from Srinivasa Ramanujan, described by Andrews (2008) as 'a self-taught Indian genius' who 'made monumental contributions to mathematics that set the stage for many of the breakthroughs in number theory in the twentieth century' (p.807). To the widely accepted result of all integers adding up to infinity, Ramanujan offered an alternative result, an exact negative fraction: $-\frac{1}{12}$. As Parker (2014) clarifies, but does not explain: 'Of course, the sum of all positive whole numbers is infinite, but if you can somehow peel that infinity back out of the way and look at what else is going on, there is a $-\frac{1}{12}$ in there' (p.297). This so-called Ramanujan summation provoked much disagreement and debate amongst mathematicians, but most of all it disrupted the comfortable expectation that intuition suffices in matters of engagement with what cannot be directly perceived. In that sense, a most psychoanalytic intervention.

This resonates with the meaning of life revealed by Douglas Adams in *Hitchhiker's Guide to the Galaxy*, as 42, a number which in ASCII code represents the wildcard *, a symbol that can represent anything (Parker, 2014). Predictability and meaning deceive.

Numbers are interesting in themselves, as sets, but also in the ways in which they interact with each other. As I indicated in the preface, my starting point in thinking about the relationship between space as unknown and the unconscious was a kind of 'automatic misreading' of Lacan's representation of the relationship between the Signifier and the signified as a 'fraction'. By virtue of many years spent thinking mathematically before encountering psychoanalysis, I found myself reading Lacan's text as one would read a new language with the accent of their mother tongue. I was interested to find that Cléro (2002) sees the bar Lacan uses to separate the signifier from the signified as gaining an increasingly marked algebraic sense in Lacan's work, as operator of the fragmentations and separations that affect the subject (p.15). For Lacan, this is the bar that turns S into $, the subject divided by the bar of language.

Given the relevance that this bar (vinculum) has to an operation that defines the experience of being of each subject of language, it is worth taking a closer look at the notion of fractions. Mathematically, division can operate as gateway to a variety of infinities built by and with fractions, some countable and some not. I recall from primary school days my own difficulty with division as an operation I perceived as radically different from the others in arithmetic – addition, subtraction, multiplication. Much later I came to understand that the latter all move between numbers of the same kind, while division is an operation that can lead to a number of a different kind from the ones one started with, in particular the move from finite numbers to numbers with infinitely many decimal places. In other words, division can operate as a gateway between radically distinct dimensions.

When two numbers are in a relation of proportionality, then the operation of division inscribes a kind of coherence and resonance between the two. When this is the case and the numerator is greater than the denominator, then the result is an integer, there is no residual. We call it a round number. For instance, $\frac{6}{3} = 2$

Most fractions create a residual, that is to say, a result that also includes an incomplete part, one that is not 'round'. For instance, $\frac{5}{2} = 2.5 = 2 + 0.5$, where 0.5 is the remainder, beyond the integer 2.

Fractions where the numerator is lower than the denominator always give a kind of incompleteness, a number that is less than one, with either finite (e.g. $\frac{1}{2} = 0.5$) or infinite number of decimal places. Fractions of the latter kind lead to a particular kind infinity, where the residual number recurs, with less and less value, but endlessly so. For instance, $\frac{13}{3} = 4.3333...$, or $\frac{1}{3} = 0.333...$, with the decimal 3 repeating to infinity.

Mathematically, resonance or proportionality is the exception rather than the rule, as irrational numbers fill densely the neighbourhood of any rational number, being most likely when the numbers involved are small (Panchelyuga and Panchelyuga, 2012). In other words, around any number that we can firmly mark on the line, there are many more numbers that cannot be inscribed in the same way. Those that cannot be inscribed are much more numerous than those that can.

Infinity as recurrence can also happen in the construction of certain fractions, for instance, of the format:

$$a_0 + \cfrac{1}{a_1 + \cfrac{1}{a_2 + \cfrac{1}{\ldots + \cfrac{1}{a_n}}}} \tag{1}$$

Such fractions can be used to generate recurrent structures known as fractals, where the same pattern is found at various scales. The mathematical formulae describing fractals appear complex, but have a simple repetitive structure and reflect both natural designs (snow-flakes, trees, clouds, broccoli) and purely abstract constructions (Falconer, 2013). This is another way of representing repetition, structurally, at various scales. What is also of interest, where fractals are concerned, is that, while geometrically they can form an infinite boundary, the area these boundaries enclose remains finite (Stewart, 2017, p.20).

A structure of this kind constitutes the foundation of Kabbalistic understanding of spiritual life. In the words of Eigen (2012), 'There is a saying, "As above, so below; as below, so above". The flow goes both ways, all ways [...] Kabbalah speaks of worlds within worlds, world after world' (p.83).

An equation is a mathematical statement about the identity of two distinct expressions on either side of an equal sign. Such a statement is also a question, in that it invites the search for the particular value that satisfies the proposed equality. When an equation can be solved, its answer is known as its solution. Some equations have more than one solution; others have none. While sometimes the solution to an equation is an irrational number, other such numbers have the property that they cannot be calculated as the solution to any equation. As Green (2008) puts it, '[S]ome numbers are more irrational than others' (p.222). Most commonly known such numbers are $\sqrt{2}$, e (the base of natural logarithms), π and τ, the golden ratio $\left(1/2\left(1+\sqrt{5}\right)\right)$. Of these, the golden ratio is the 'most irrational' because the best rational approximations to it approach it slowly. Also, e and π are also known to be transcendental numbers, that is to say, not the solution of any polynomial equation with integer coefficients of the format $a_n x^n + a_{n-1} x^{n-1} + \ldots + a_2 x^2 + a_1 x + a_0$. In other words, they exist without being the answer to a question that can be formulated algebraically. This goes to the heart of the notion of symbolisation and mathematical impossibility: it is not possible to construct a square which has the same surface as a circle (squaring the circle, as it is known). '[T]he area of the simplest of all geometric figures, the circle, cannot be determined by finite (Euclidean) means' (Kasner and Newman, 1940, p.80). Interestingly, in the entire universe of numbers, transcendental numbers are thought to be the most numerous, whilst at the same time there being no proof that such a number exists. These are numbers that 'lie beyond those that arise through Euclidean geometry and ordinary algebraic equations' (Higgins, 2011, p.90), filling in the space between known numbers like a kind of 'dark matter of the number world' (Higgins, 2011, p.91). In the striking words of Marcus (2014), most real numbers lack representation; they exist in a state of 'joined ownership, as if glued to each-other [*my translation*]' (p.33), as most of what is known about numbers is accessible at a global rather than at an individual level. For most numbers, the further one goes down the sequence of decimal places, the less possible it becomes to distinguish one number from the next, to distinguish what belongs to which number[2]. 'Real numbers are, in general, known through approximate values, therefore through metonymical processes' (Marcus, 2014, p.33). This notion of circumscribing something that escapes transcription through metonymy is akin to the approximation of desire in the metonymical movement of demand, a Lacanian idea to which I return in Part III.

Like the unconscious, desire and its generation are not knowable directly, and nestle in the gaps of what we think we know, like the book in Borges' story, disrupting the continuity and order of our subjective experience.

So, despite our well-rooted familiarity with integers and fractions, it is not such numbers that make up most of the space of numbers, but irrational ones, each of them containing in themselves a fragment of infinity. Thus, while there are an infinity of rational numbers, these occupy no space at all on the number line. Instead, this space is taken up by a larger infinite set (in the way Cantor helped distinguish between countable and uncountable infinities, some smaller and some larger), that of irrationals: 'Even though there are rational numbers everywhere on the number line, they take up no space at all. If we were to throw a dart at the number line, it would *never* hit a rational number' (Seife, 2000, p.156). Uncountable infinities are non-denumerable; they are too big to be counted by the (infinitely large) class of integers, with which they cannot be put into a one-to-one correspondence. By *reductio ad absurdum*, Cantor showed this to be the case for the class of real numbers between 0 and 1 – and therefore for all real numbers (Kasner and Newman, 1940, p.50). Not all infinities are the same.

As we have seen, most numbers, those falling under the umbrella of real numbers, are mathematically thought about as positions, as points along one line, so therefore as inhabiting a unidimensional space. Imaginary and complex numbers, on the other hand, inhabit a different kind of space, a two-dimensional one. Any point in this space can be located through a pair of coordinates: one real and one imaginary akin to the notions of longitude and latitude on a map. The real number coordinate is a position on a number line of the kind we have considered so far. Imaginary numbers are those involving $i = \sqrt[2]{-1}$, and they specify a position in a new, posited dimension (more on this in the next section). The sense of the two terms is specific to mathematics, close to the ordinary sense of these words, but different from the meaning that Lacan came to attribute to them. This can be misleading both in mathematics[3] and in Lacanian speak as in both cases the ordinary meaning of familiar words is discarded in favour of complex and precise technical connotations.

The relationship between these varied sets of numbers can be represented as is illustrated in Figure 3.2.

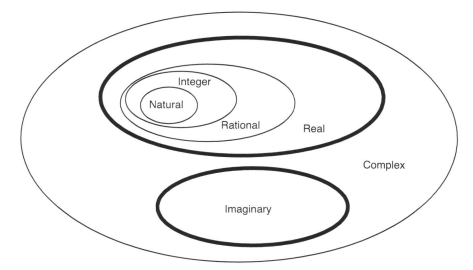

Figure 3.2 Number sets.

Like any picture, this representation is deceiving. Although this is a two-dimensional map, with the exception of complex numbers, which occupy a two-dimensional space, all the other types of numbers are located on a single line. Also, the relative size of each set as represented by an elliptical disk here is not to scale. Specifically, the numbers we can name and represent in this way are the minority. Indeed, most of the universe of numbers is made up of numbers that cannot be computed in any way, or of dark numbers, as Parker calls them:

> not only are the whole numbers and rationals we use in everyday life as insignificant part of the total sea of numbers but, even when they are combined with algebraic and transcendental numbers, we are still only on the surface. Surrounding us is an unimaginable volume of uncomputable numbers: the dark numbers which we know are there but which we cannot grasp.
> (Parker, 2014, p.382)

In this interplay between notions of dis/continuity and infinity, questions persist. Is there a number at every point on the number line, or are there any gaps? Is the line a continuous space, or is it somehow porous? Can we still refer to it as a line, if that is the case? Second, are there numbers that cannot be positioned on the line? What kind of space do they inhabit?

On the first two questions, which so much preoccupied Cantor, the consensus remains that the line is densely populated, and that between any two numbers there is an infinity of other numbers, such as in the space between 0 and 1.[4]

As for the remaining questions, the answer is given by higher-dimensional numbers, to which I turn in more detail in the next section.

3.2 From numbers to posited dimensions

Equations do not always have a solution, or, if they do, they may be of a rather complicated kind. For instance, some equations do not have solutions from the unidimensional realm of the real number line, but they do from a two-dimensional posited space. One particular way of classifying numbers is in terms of whether they represent the solution to any equation. We saw that transcendental numbers are irrational numbers that do not form the solution to any equation, but they exist somehow as answers without a question.

Imaginary and complex numbers also evolved in relationship to equations, but they came about as a way of dealing with those equations for which real solutions were not available, as they involved the square root of a negative number. As any real number, whether positive or negative, becomes a positive number when squared (i.e. $a \times a = a^2 = (-a) \times (-a)$), the only way out was to *suppose* the existence of a number which, when multiplied with itself, would become a negative number, namely, $i \times i = i^2 = -1$. In this way the square root of any negative number could be seen as the product of a real, familiar side, and a posited, imaginary side. For instance, $\sqrt{-5} = \sqrt{-1} \times \sqrt{5} = \sqrt{5} \times i$

This supposition, not unlike the supposition of the unconscious, introduced a new dimension and thus made it possible to enter a space of new possibilities. The introduction of imaginary numbers transformed the space of numbers from the number line to the complex plane, opening access to a second dimension. In this new space, each number is defined by two coordinates: a real one (along the horizontal axis) and an imaginary one (along the vertical axis). Numbers that have both a real (*Re*) and an imaginary (*Im*) component are called complex numbers and they sit in the space opened up by the two axes rather than on any of them (see Figure 3.3). With the addition of the vertical axis, movement in the space of numbers became possible in a dimension hitherto unavailable. As Moncayo and Romanowicz (2015) emphasise, the imaginary axis creates 'a second dimension different than the one we know with real numbers' (p.123). Likewise with the unconscious, whose supposition transforms the space of the psyche from a knowable, visible, conscious, measurable domain into something more complex, a new space that allows for the unknown and for operations which make possible transformations in the way a life is lived. The analytic interventions of the clinic, through speech and the analytic act, operate within and according to the laws of this new dimension, with real effects, as we will see in Part IV.

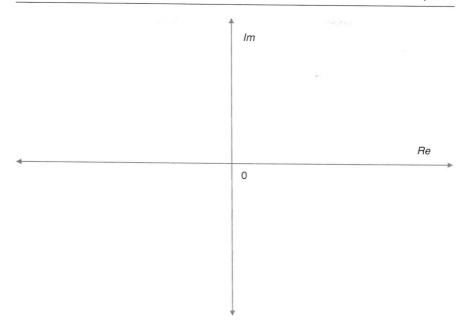

Figure 3.3 The complex plane.

Both in the world of numbers and in the clinic, one departs from what is immediately apparent, traversing, by symbolic means, a posited domain, only to return to the familiarity of daily life and find it subtly but fundamentally transformed by this process. As far as complex numbers are concerned, it is important to note that both dimensions start from zero, which is all they have in common (see Figure 3.3). Zero, a particular kind of nothing, operates as the entry point to each of them. Also, in many of the equations that become solvable once complex numbers are allowed for, the imaginary part appears only then to disappear at some part in the calculation, leaving solutions that only contain a real component. It is as is some transformations require movement though this additional and radically different dimension accessible via the imaginary axis, away from and then back to real numbers, a kind of detour through a higher dimensionality.

A real number is just a point on the horizontal line, as an imaginary number can only occupy a position on the vertical axis, while a complex number sits outside both axes, in the plane opened up by the two. For instance, 3 and $5 + 2i$ has the real component of 5, the imaginary component of $5i$, and is located as illustrated in Figure 3.4.

Note that the posited imaginary number i, the square root of minus one, operates as unit of measurement on the newly created axis.

Complex numbers do not just exist in two-dimensional space, but also in four-dimensional and eight-dimensional space. These are the so-called quaternions, first identified by William Hamilton (1866), and octonions (part of the wider class of

52 The unconscious as inaccessible space

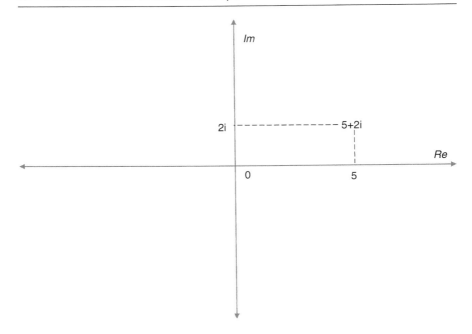

Figure 3.4 Mapping of numbers in the complex plane.

so-called hypercomplex numbers). Quaternions are used to represent movement on toric trajectories (Banchoff, 1996, p.182), to which we return in Part III. Interestingly, no equivalent three-dimensional numbers exist; neither are there any such numbers in any other dimensions (Parker, 2014, pp.394–396). It is important to recognise that such numbers are constructed not only on the supposition of the imaginary domain, but also on that of dimensions higher than the three we can perceive. Yet the construction of both quaternions and octonions has allowed significant progress in physics, in particular in understanding and addressing practical questions about waves, heat, electricity, magnetism (Stewart, 2013), as well as in the fields of string theory, special relativity and quantum logic, or in the more familiar realm of computer graphics (Higgins, 2011, p.122). What this shows, among other things, is that a supposition is not sufficient, that one cannot just make things up and that certain constraints regarding consistency and rigour on the properties of what is supposed to be define the outcome and the impact possible, always ultimately verifiable at the level of what can be observed.

Returning for a moment to transcendental numbers, one relationship that brings together numbers from these different registers and which deserves a special mention here is Euler's equation:

$$e^{i\pi} + 1 = 0 \qquad (2)$$

where e is the base of natural logarithms (2.71828…) and π is most widely known as the ratio of a circle's circumference to its diameter (3.14159…).

This equation is only a statement, not a question, as it contains no unknown. Here, e and π belong to the realm of irrational numbers, and have an infinite number of decimal places, while i is the square root of -1. Spatially, of all the components of equation (2), only zero and one can be placed on a system of Cartesian coordinates[5] with any degree of precision. In Lacanian terms, this equation is an encounter with the Symbolic in pure form: although referred to by many as a thing of beauty, this identity is not one that can be reduced to the Imaginary in the Lacanian sense. In the words of Benjamin Peirce, a philosopher and mathematician whose work was familiar to Lacan, 'it is absolutely paradoxical; we cannot understand it, and we don't know what it means. But we have proved it, and therefore we know it must be the truth' (Peirce, cited in Kasner and Newman, 1940, p.104).

However, being able to locate zero and one is not a small matter, given that, as a topological space, the entirety of the real line we examined above is homeomorphic[6] to the open interval (0, 1). So, the defining infinity of the space of real numbers can nestle in the space between nothing and the first mark of a whole entity, but where neither zero nor one is included. In the words of Levinas, 'Infinity is the adequate measure of all that is: it is the finite straight line that is in potentiality and the infinite straight line in act, actualizing that which was only potential in the finite straight line' (Levinas, 1999, p.66). This particular relationship between the finite and the infinite is highly relevant, given that, from a Kantian perspective of describing the finite independently of the infinite (Levinas, 1999, p.72), the distinction at play here is resonant of the radical differentiation between consciousness and the Freudian unconscious: the infinite is not just more of the finite, in the same way that the unconscious is not just some hidden consciousness.

We saw how number theory maps out with great clarity the interplay between that which can be known and symbolised and that which escapes, and between the finite and the infinite. We also made a start on the question of dimensionality and the possibilities of movement that a new dimension opens up in ways that contrast with the impossibilities that pervade a space where this extra dimension is not accessible. We now turn to the more specific links between these mathematical concepts and their counterparts in the unconscious.

3.3 Infinity and incompleteness in the unconscious

As with many of the most sophisticated uses of mathematical thought in the field of psychoanalysis, the main contributions in the use of elements of number theory belong to Lacan and his followers.

Whilst we all use numbers in our daily life, few of us could actually provide a suitable definition. Despite their ubiquity in various aspects of the quotidian, it is important to remember that numbers are purely abstract constructions, without any equivalent in reality. At the same time, each number is unique, and the question of location or position is utterly relevant, not only in the case of digits (2019 is not the same with 2091), but also in relationship to other numbers. In this regard, what is of interest is not just any particular number in itself, but the system they create, namely, a set of entities on which operations can be performed (Gouvêa, 2008, p.82), a space where transformation is possible and something new can be generated.

Thus, Lacan made much use of numbers as part of his exploration of the power and the limitations of the register of the Symbolic. It is important to keep in mind that mathematical formalism itself developed starting with number theory, so the relevance of numbers in this context is at least two-fold: in terms of their capacity to engage with something about the Real and in terms of their power of development of a purely Symbolic register. As Rosset (2013) puts it, for Lacan, mathematics represented the royal road for accessing the Real of structure (p.94). As far as the concept of the unconscious itself is concerned, Lacan expresses the process of approaching it in terms of the nature of infinitesimal calculus, whereby 'it is only by a leap, a passage to the limit, that it manages to realise itself' (Lacan, 1979, p.19). In his earlier work, Lacan's elaboration of the Symbolic relied primarily on his take on notions of linguistics, and placed much faith in the view that the linguistic structure is the one 'that assures us that there is, beneath the term unconscious, something definable, accessible and objectifiable' (Lacan, 1979, p.21). However, as his thinking developed, the emphasis shifted from a question of accessibility to one of impossibility, and the mathematical underpinnings of his thought also shifted from calculus to topology. More on this in Part III.

Much of Lacan's early interest in numbers related to the idea of repetition and of counting. As he was formulating his conception of the unconscious structured like a language, centred on the place of the signifier in both dividing and representing the subject, Lacan placed great importance on early numeral systems of representation – the marking of strokes on wood or bone, each presumed by him to represent the act of killing an animal – which he linked fundamentally to being human as defined by accessing the Symbolic realm (Lacan, 2020, pp.56–57). Psychoanalytically, he linked this basic mark to the concept of unary trait which he explored at length in this seminar, and which in his view underpins structurally the mechanism of **repetition** defining of the operation of the psyche. This trait is 'the mark of absolute difference' (Moncayo and Romanowicz, 2015, p.131).

Thus, although counting is about repeating instances of 'one', creating a sequence of evenly spaced marks, this operation of repetition both hides and reveals the fact that each instance is both uniquely different and the same with all the others.

Lacan also linked counting to **identification** and to the nature of the signifier which makes language work not in its relation to a meaning or signified, but in relation to other signifiers, taking its value not absolutely but relatively, operating through difference. Like the markings on the bone, each signifier is a one, but they are not the same 'one'. They are the same, but also different in their repetition. As Chemama and Vandermersch (2009) stress, this difference of the signifier from itself in its repetition is considered by Lacan one of its fundamental properties (p.584). If we consider this in conjunction with his view that the unconscious is structured like a language, then it is possible to understand that repetition is a structural feature. Also, the concept of unary trait makes it possible to see that repetition in the Freudian sense is not an eternal return to a fixed point. Instead, the recurring instances are both the same and distinct. It is worth noting that Lacan developed this concept at the same time as he was working on the topological properties of surfaces other than the sphere, which he rejected as 'the intuitive and mental model of the structure of the cosmos' (Lacan, 2020, p.210), and which he described as apparently unified and 'conflict-free' (see Greenshields, 2017, p 39). Across this exploration of the relevance of topology to the structure of both language and the unconscious, the torus in particular yielded fresh insights in terms of the landscape of the interplay between repetition, demand and desire. As we will see in Part III, movement on such a surface can entail not only repetition, but also the possibility of access to pathways that are not defined by it. Also, as we will see in the next chapter, repetition can be understood spatially in terms of dimensions, namely, not as discrete encounters with a sequence of similar things, but as a recurrence of encounters with various instances of one single entity. In each encounter, the subject is like one of the blind men who touch and describe various parts of the same elephant, failing to arrive, both individually and as a group, at a view of the totality of their experience of an object they cannot apprehend directly in its totality. That particular, partial and fragmented understanding indexes a gap in the form of inaccessibility. For a human being whose physical life is defined in three dimensions, such inaccessibility is no different from an impossibility, the kind of gap best captured structurally by topology, as we will see in Part III.

Besides counting as a repetition of unary inscriptions, Lacan also devoted attention to irrational numbers and their properties, focused as he was on their resonance with aspects of the subjective experience that, whilst escaping visual representation, could be considered and expressed only in Symbolic terms. Thus, he devoted considerable attention to the golden ratio and its properties (Moncayo and Romanowicz, 2015), in particular in Seminar XIV of 1966–1967, in his elaboration of the concept of the **phallus**. As the authors stress, the motivation for this pursuit is fundamentally psychoanalytic as '[A]lthough numbers follow the rules of knowledge what they produce is not knowledge but truth' (p.150). The same can be said about the work of analysis.

Of particular interest amongst irrationals is the square root of 2, $\sqrt{2}$, which Pythagoras's theorem showed to be the length of the diagonal for a square with side length of 1. The length of the side and of the diagonal are 'fundamentally incompatible, or incommensurable' (Higgins, 2011, p.76), as these are numbers radically different from each other in their properties. According to legend, Hippasus, the man who was first to work out this result, was thrown overboard from a boat he was travelling in, so unpalatable were his findings with his contemporaries (Stewart, 2017, p.25).

Although a square and its diagonal can be drawn, the measure of the sides and of the diagonal, respectively, belong to different realms of the world of numbers. As Lacan puts it in the early 1950s in *Seminar II*, 'There is no common measure between the square's diagonal and its side' (Lacan, 1991, p.256). This is of the same nature with the lack of accord (in the grammatical sense) between the body and the unconscious that he examines some two decades later in Seminar XXII, *R S I* (1974–1975). The nature of this radical incompatibility pertains to the difference between distinct spatial dimensions: one cannot be turned into the other; they merely coexist, both separately and together.

In his earlier work, Lacan examined the properties of $\sqrt{2}$ as a means of punctuating something about the distinction between the Imaginary and the Symbolic domains, while warning against the confounding of the two registers. In this sense, one could argue that Lacan treated the registers as distinct dimensions in the spatial sense. In the second session of *Seminar II*, he brought to the attention of his audience one of Plato's earliest dialogues, the Meno dialogue, which centres on a mathematical riddle related to the one solved by Oedipus. The slave in the dialogue is asked to double the surface of a square. The key to being able to do this is to grasp something about the relationship between the sides and the diagonal of the square. As Lacan emphasises, there is 'a fault-line between the intuitive element and the symbolic element', and 'here we put our finger on the cleavage between the imaginary, or intuitive, plane [...] and the symbolic function which isn't at all homogeneous with it, and whose introduction into reality constitutes a forcing' (Lacan, 1991, p.18). The two relate to each other in a way that dimensions do. One cannot more from one to the other through more of the same. Instead, the move is a radical one, a sudden opening to a space hitherto unavailable.

In the more complex version of this puzzle, when asked for a way to banish the plague of Athens, the Oracle of god Apollo at Delos set Athenians the task of doubling the volume of an altar which was a perfect cube. This Delian Problem (Wells, 1991, p.49) is the three-dimensional equivalent of the problem posed to the slave in Meno's dialogue. The task was an impossibility in terms of what could be accomplished with rulers and compasses alone. In other words, there is no route that goes from $\sqrt{2}$ to $\sqrt[3]{2}$ (Higgins, 2011, p.78), and the solution could not be attained by Euclidean means (although Wells, 1991, p.50, offers a Euclidean approximation). Likewise, the move from 1 to $\sqrt{2}$ requires more than mere repetition.

The extent to which such a variety of numbers, all real, can or cannot be located easily in the particular space of the number line hints at a notion of discord (again, in the grammatical sense), of something belonging to radically different domains. In the puzzle that Oedipus solved, both $\sqrt{2}$ and $\sqrt[3]{2}$ are numbers, but one cannot move from one to the other by counting; the space between them cannot be traversed by mere repetition, but it requires access to some other dimension. It is as if the path that links the two strays off the number line into another dimension, only to return to it at some other place.

As we have seen, the introduction of imaginary and complex numbers was a radical departure in the evolution of mathematical thinking about numbers, in that, through a supposition, a new dimension, hitherto unavailable, became accessible and new constructions with real and immediate implications became possible.

Lacan's particular use of complex numbers has generated numerous responses, from furious attacks to passionate defences, perhaps precisely because he transgressed, as he often did, in so many ways, the grid of established meanings.

Lacan concentrated on these as he focused on articulating something about what remains inaccessible and unavailable to representation either on the Imaginary (in the Lacanian rather than the mathematical sense) or the Symbolic planes, that is, the phallus. He conceived of it as analogous to the square root of -1, something posited but not open to observation or experience, something ultimately inaccessible (Plotnitsky, 2009). The question of inaccessibility is fundamental to the very notion of spatial dimensions. For instance, movement in the space of real numbers is only possible along the number line, back and forth, or left and right, and nowhere else (Figure 3.1). Mathematically, the move from real to complex numbers is a move to an increased dimensionality: from representing each number as one point on a line to representations in the plane. The dimension introduced by the imaginary axis addresses the impossibility that remaining in the one-dimensional space of the line number produces. What cannot be reached in one dimension becomes accessible when a new dimension is added.

In a detailed psychoanalytic examination of numbers in relation to the Lacanian Real, Moncayo and Romanowicz (2015) allude to the question of dimensions and their relevance with regard to experience. Thus, quite early on, they specify:

> So when we say, 'I think therefore I am' we can be correct only if we suppose that these two things occur in different dimensions: the dimension of the ego and the dimension of the unconscious (the core of being according to Freud).
>
> (p.6)

More importantly, they regard the interlinking of the three registers, in what Lacan called the Borromean knot, as the interplay of dimensions of experience that interact and intersect one another (Moncayo and Romanowicz, 2015, p.9). They spell out the rather cryptic Lacanian names of the Symbolic, Imaginary and Real as 'three dimensions: name, image, and beyond name and image' (p.18). Bursztein (2008) insists on the symptom as a fourth component of the unconscious psychic knot (p.25). More on this in Part III.

Moncayo and Romanowicz (2015) also warn against the loss of consistency in this structure when dimensions are collapsed '[I]n an attempt to reduce the contradictions and inconsistencies between the dimensions of the knot' (pp.10–11). This resonates with the early Lacanian criticism of the psychoanalytic approach prevalent in the 1950s, where what he untangled as the Symbolic and Imaginary registers would be collapsed into one, and therapy reduced to an interaction defined by the Imaginary axis in the L-schema (Lacan, 1955, 1991).

As illustrated in Figure 3.2, the various sets of numbers are largely conceived of in a nested structure, with each wider set incorporating smaller sets as well as a domain that is specific to them and to higher sets. This quasi-nested setup has a correspondence to the registers of experience explored by Lacan: integers, and the illusion of completeness they offer, belong to the (Lacanian) Imaginary; complex numbers with rational coefficients establish, like the Symbolic, positions relative to a system of axes, while allowing for the same point to represent something else in another system of axes. Finally, irrational numbers with their infinite decimal places can be approximately rounded to fit into the other two registers, but something always escapes – the remainder of the Real. As Bursztein (2008) elaborates,

> 'The subject is not only divided but always dividing himself between the structure of the identifications inferred by the line of signifiers of the unconscious chain and what does not pass into the signifying net of language. This remainder, which constitutes the heart of our being, our immovable core, constantly boosts all unconscious processes.
>
> (p.27)

In his later work Lacan addresses the question of dimensions more directly, albeit briefly. As he introduces various representations of the Borromean knot, whose structure he proposes as defining the relative positions of the three registers of experience (Imaginary, Symbolic, Real), Lacan presents the nature of the encounter between spatial dimensions in terms of their commonality, as represented by the way the lower dimension intersects or cuts the higher one:

> What cuts a line is a point. Since a point has zero dimensions, a line is defined as having one dimension. Since what a line cuts is a surface, a surface is defined as having two dimensions. Since what a surface cuts is space, space has three dimensions.
>
> (Lacan, 1998, p.122)

He leans heavily here on Poincaré's definition. Also, what he calls space here is only three-dimensional space. The question to be considered is what happens with the move to a four-dimensional space, and this is what we are going to examine in the next chapter.

3.4 The unconscious as inaccessible space between points of encounter

Although we can all count, we might not immediately see that counting amounts to putting two classes into one-to-one correspondence with each other: the class of what is counted and the class of integer numbers.

Since Cantor, mathematicians came to distinguish between countable and uncountable infinities, on either side of zero. Also, since Cantor, it is known that the set of all real numbers is not a countable infinity (Dauben, 2008, p.779). In other words, there is no way of knowing what happens at every point in the line-space inhabited by numbers. It is not that there are gaps in the line (although Cantor was much troubled by this possibility, which he considered in detail), but that some points in the space of numbers cannot be known or accessed, but merely posited, just like the unconscious.

Attributing this positioning to Freud himself, Lacan explicitly places the unconscious in the gap between cause and effect, where something 'of the order of the non-realized' (Lacan, 1979, p.22) is located. 'At first, the unconscious is manifested to us as something that holds itself in suspense in the area [...] of the unborn' (Lacan, 1979, p.23). He recognises this space explicitly as a dimension which is not unreal but unrealised (Lacan, 1979, p.23), and links this to Freud's notion of the navel of the dreams, which designates the ultimately unknown centre of dreams and which is ultimately a gap not to be stitched up, but opened with care (Lacan, 1979, p.23). This formulation of the unconscious as gap already puts into question the notion of inaccessibility in favour of that of impossibility, which we examine in Part III.

The question of the spaces between what can be signified and what might reside in them was one that Lacan did consider in some detail in his formulation of the unconscious structured like a language, where the subject's being happens in the movement from one signifier to the next. If we understand spatially the dimension he thus introduces, the so-called *dit-mension* (Lacan, 1972), the residence of the spoken, then language marks a sequence of points of encounter with another dimension, defined by a continuity that escapes and can only be perceived as discontinuity. As he put it Seminar XI, the praxis of psychoanalysis is 'a concerted human action [...] which places man in a position to treat the real by the symbolic' (Lacan, 1979, p.6). The continuity that belongs to this other dimension can only be articulated as and through points of discontinuity, with language as the locus of those points. As we will see in the next chapter, it is possible to formulate a spatial interpretation of the notion of the unconscious structured as a language. There is no false unity in the unconscious but only split and rupture (Lacan, 1979, p.26). Thus, our encounters with any unconscious formation unsettle us as an encounter with discontinuity. As Marcus (2014) underlines, with reference to the work of René Thom, the human tendency is to perceive continuity (which is akin to infinity), but to comprehend only what is finite (p.11).

As we saw with complex numbers, the positing of imaginary numbers made it possible to access a new dimension, away from the spaces between what could be inscribed on the real number line alone. Likewise, positing the existence of a higher spatial dimension can reconcile what appears fragmented in three dimensions with the compactness and continuity of a space defined by four dimensions, in a way that corresponds to the relationship between the fragmented and incomplete Symbolic and the compact, 'lacking in lack' notion of the Real introduced by Lacan.

In other words, the gaps and discontinuities that we perceive in the daily encounters with the unconscious (dreams, slips of the tongue, bungled actions, etc.) can be understood as markers of a space whose continuity is determined by a further dimension which we cannot access directly, but only through a sequence of such discrete (as opposed to continuous) points of encounter.

Thus, what this examination of the numbers space has shown is that even spaces conceived of as continuous, in which every point can be uniquely located, and where distances can be measured accurately, are territories where a notion of discontinuity persists, where what can be expressed and inscribed is only a point of discontinuity in a continuum located elsewhere which remains largely unknown and inaccessible.

Many arrive to analysis in search for an answer to their suffering, The Answer whose existence seems vaguely possible but out of reach, just like the navel of the dream. This starting point rides on an implicit assumption of the unconscious as infinity mis/understood as a place which could be, at least in principle, approached by a sequence of moves (or sessions, or books of self-help). Alongside this, there is also the implicit assumption that the unconscious is a kind of space of higher dimensionality, inasmuch as all that is apprehended about it as part of our being and its pain arrives somehow incoherent and incomplete, with its completeness in some kind of elsewhere. To this, we turn in the next chapter.

The realisation that The Answer might not exist at all takes shape only gradually in the work of analysis, through which one finds a way not to deny or resolve the impossibility, but to figure out a relative and highly personal way with life in a world marked by an absolute lack.

Notes

1 In most general terms, a limit of a convergent sequence is a value to which a convergent sequence tends to, a value to which it gets gradually closer.
2 This is a clarification I owe to Serban Sovaiala.
3 Kasner and Newman (1940) define mathematics as 'the science which uses easy words for hard ideas' (p.4).
4 Indeed, an alternative way of thinking about numbers has been introduced in the late 1960s by John Conway, who developed the notion of surreal numbers. These are closer to any real number than any other neighbouring real number (Conway, 1995).
5 Introduced by Descartes, this is similar to the system of coordinates depicted in Figure 3.3, except that in the original Cartesian version, both axes represent real numbers. Each point in the plane is uniquely defined by a pair of coordinates, read in relation to the point of intersection, called the origin, conventionally set at zero, which has coordinates (0,0).
6 Homeomorphic spaces are the same from a topological point of view. Technically, each point in one space corresponds to one and only one point in the other, and two neighbouring points in one space correspond to a unique pair of neighbouring points in the other (Nasio, 2010, p.29).

References

Adams, C.C. (2000). *The knot book: An elementary introduction to the mathematical theory of knots*. Providence, RI: American Mathematical Society.
Andrews, G. (2008). Srinivasa Ramanujan. In Gowers, T. (ed.). *The Princeton companion to mathematics*. Oxford: Princeton University Press, pp.807–808.
Armstrong, M.A. (1983). *Basic topology*. New York: Springer-Verlag.
Arnold, B.H. (2011). *Intuitive concepts in elementary topology*. New York: Dover Publications.
Banchoff, T. (1996). *La quatrième dimension: Voyage dans les dimensions supérieures*. Paris: Pour La Science Diffusion Belin.
Borges, J.L. (1998). *The book of Sand and Shakespeare's Memory*. London: Penguin Books, pp.89–93.

Bursztein, J.-G. (2008). *On the difference between psychoanalysis and psychotherapy*. Paris: Nouvelles Etudes Freudiennes.

Charraud, N. (1994). *Infini et inconscient: Essai sur Georg Cantor*. Paris: Anthropps.

Chemama, R. and Vandermersch, B. (2009). *Dictionnaire de la psychanalyse*. Paris: Larousse.

Cléro, J.-P. (2002). *Le vocabulaire de Lacan*. Paris: Ellipses.

Conway, J. and Guy, R.K. (1995). *The book of numbers*. New York: Copenicus.

Dauben, J.W. (2008). Georg Cantor. In Gowers, T. (ed.). *The Princeton companion to mathematics*. Oxford: Princeton University Press, pp.778–780.

Eigen, M. (2012). *Kabbalah and psychoanalysis*. London: Karnac.

Falconer, K. (2013). *Fractals: A very short introduction*. Oxford: Oxford University Press.

Gouvêa, F.O. (2008). From numbers to number systems. In Gowers, T. (ed.). *The Princeton companion to mathematics*. Oxford: Princeton University Press, pp.77–82.

Gowers, T. (2008) (ed.) *The Princeton companion to mathematics*. Oxford: Princeton University Press.

Green, B. (2008). Irrational and transcendental numbers. In Gowers, T. (ed.). *The Princeton companion to mathematics*. Oxford: Princeton University Press, pp.222–223.

Greenshields, W. (2017). *Writing the structures of the subject: Lacan and topology*. London: Palgrave Macmillan.

Hamilton, W. (1866). On quaternions. In Smith, D.E. (ed.). (1959). *A source book in mathematics*. New York: Dover Publications, pp.677–683.

Higgins, P.M. (2011) *Numbers: A very short introduction*. Oxford: Oxford University Press.

Kasner, E. and Newman, J. (1940). *Mathematics and the imagination*. New York: Dover Publications.

Lacan, J. (1955). Seminar on "The purloined letter". In Fink, B. (ed.). (2002). *Écrits*. London: WW Norton, pp.6–48.

Lacan, J. (1972). L'Étourdit. In Lacan, J. (2001). *Autres Écrits*. Paris: Éditions du Seuil.

Lacan, J. (1979[1973]). *The four fundamental concepts of psychoanalysis*. London: Penguin Books.

Lacan, J. (1991[1978]). *The seminar. Book II. The ego in Freud's theory and in the technique of psychoanalysis, 1954–1955*. London: WW Norton.

Lacan, J. (1998[1975]) *The seminar. Book XX. On feminine sexuality, the limits of love and knowledge, 1972–1973*. London: WW Norton.

Lacan, J. (2020). *L'identification: Séminaire 1961–1962*. Éditions de l'Association Lacanienne Internationale. Publication hors commerce.

Levinas, E. (1999). *Alterity and transcendence*. London: Athlone Press.

Marcus, S. (2014). *Singurătatea matematicianului: discurs Academia Română, București 2008*. Bucharest: Spandugino.

Matte Blanco, I. (1988). *Thinking, feeling, and being: Clinical reflections on the fundamental antinomy of human beings and world*. London: Routledge.

Moncayo, R. and Romanowicz, M. (2015). *The real jouissance of uncountable numbers: The philosophy of science within Lacanian psychoanalysis*. London: Karnac.

Nasio, J.-D. (2010). *Introduction à la topologie de Lacan*. Paris: Petit Bibliothèque Payot.

Newton, I. (1666). *Descartes, space and body*. https://www.earlymoderntexts.com/assets/pdfs/newton1666.pdf.

Panchelyuga, V.A. and Panchelyuga, M. (2012). Resonance and fractals on the real numbers set. *Progress in physics*, 4, pp.48–53.

Parker, M. (2014). *Things to make and do in the fourth dimension*. London: Penguin Books.
Plotnitsky, A. (2009). On Lacan and Mathematics, *Oeuvres & Critiques*, XXXIV(2), pp.143–162.
Poincaré, H. (1912). Pourqui l'espace a trois dimensions. *Revue de métaphysique et de morale*, 20(4), pp.483–504.
Rosset, J.P. (2013). Le réel entre mathématiques et psychoanalyse. *Seminaire de psychanalyse 2012–2013*, ALI Alpes-Maritimes-AEFL, pp.83–95.
Seife, C. (2000). *Zero: The biography of a dangerous idea*. Chippenham: Souvenir Press.
Smith, D.E. (ed.). (1959). *A source book in mathematics*. New York: Dover Publications.
Stewart, I. (2001). *Flatterland: Like Flatland, only more so*. New York: Basic Books.
Stewart, I. (2013). *Seventeen equations that changed the world*. London: Profile Books.
Stewart, I. (2017). *Infinity: A very short introduction*. Oxford: Oxford University Press.
Vappereau, J.M. (2006). La D.I. http://jeanmichel.vappereau.free.fr/textes/La%20DI.pdf.
Vivier, L. (2004). *La topologie: L'infini matrisé*. Paris: Le Pommier.
Wells, D. (1991). *The Penguin dictionary of curious and interesting geometry*. London: Penguin Books.

Chapter 4

The rigour of spatial dimensions – of shadows and recurrences

While human vision is limited to the registration of two-dimensional representations, the body moves through three-dimensional space. Mathematically, we can conceptualise lower dimensions, as well as higher ones, yet we have no direct access through our senses to more than three spatial dimensions. Nevertheless, something in the nature of a mediated access happens through the unconscious, not only in dreams – with their complex and gravity-defying features, with their absence of contradiction and disregard to the familiar order of waking life – but also in the daily occurrence of repetitions and overlaps that we can at best recognise, but only with great difficulty, and, after much psychoanalytic work, can engage with in a way that can lead to change. One could say that a lot of things do change through analysis, but, in a specific sense, only one thing changes, namely, the position of the subject within the space of their own unconscious. It is the question of the dimensionality of the unconscious as space that is of interest here.

In order to develop an understanding of the unconscious as space, in the mathematical sense, it is essential to consider the notion of dimension. This is a term often used in daily life, often metaphorically, yet it has its specific and complex signification, mathematically. Like many other key notions in mathematics, dimensionality was for a long time taken as implicitly given, and became formalised explicitly only relatively recently.

Whether it can be perceived by the senses or not, any space is defined by its dimensionality. The main thesis put forward here is that the unconscious can be understood as a four-dimensional space, which means that, by necessity, it can never be apprehended in its totality, but only in a seemingly fragmented way, through the means of perception we have available for life in what we experience as three-dimensional physical space. Mathematically, a four-dimensional space includes the entirety of any three-dimensional space, and something beyond. What appears to be fragmented and repetitive at the level of experience can be understood as the necessary consequence of dimensional incompatibility between our capacity to perceive and experience, and that which is to be perceived as arriving from the unconscious, given that something four-dimensional would not fit into three-dimensional space without fragmentation, overlap or repetition.

The Euclidean conception of space, rooted in classical Greek geometry, in particular Euclid's axioms and postulates, has dominated for the last 2,000 years or so.[1] Euclidean spaces are smooth and continuous worlds defined by straight lines, where positions are defined by a number of parameters (coordinates) equal to the number of dimensions of the space. Thus, in one dimension, one parameter is needed to locate a point. For instance, taking strictly the line formed by the edge of a ruler, each point alongside it is uniquely defined, e.g. the point marked 7 cm is uniquely and unequivocally identified. In two dimensions, two parameters are required to determine the position of an element in it (e.g. latitude and longitude on a map), while in three dimensions, the location of a point is defined by three parameters (e.g. geographical location including altitude). Euclid's definitions were brief: a line is what extends one way, a plane that which extends two ways and a body that which extends three ways (Crilly, 1999, p.2). The Euclidean view of space takes the concept of dimension as given and operates exclusively in up to three dimensions. Despite the intuitive nature of this approach to space, it is important to remember that spatial dimensions are a construct rather than an intrinsic feature of the world we inhabit. Nevertheless, as Charles Hinton,[2] the British mathematician contemporaneous with Freud who devoted most of his work to the study of the fourth dimension, reminds us, '[D]imensions are artificial enough, but in the multiplicity of them we catch some breath of nature' (Hinton, 1901, p.121).

Developed relatively recently, various departures from this prevailing Euclidean view of space were radically different, as they moved away from the accepted view towards something hitherto inconceivable. Kasner and Newman (1940) argue that '[T]here has been no more sweeping movement in the entire history of science than the development of non-Euclidean geometry' as a challenge to Euclid's view of space which 'stood for more than twenty centuries in lone, resplendent and irreproachable majesty' (p.134).

The earliest formal developments in mathematics with regard to defining spatial dimensions were, more or less contemporaneous with Freud. Until the publication of the paper by Möbius in 1827, *On higher space*, all reference to geometry in spaces of more than three dimensions were, as Smith (1959) puts it, 'in the form of single sentences pointing out that we cannot go beyond a certain point in some process because there is no space of more than three dimensions' (p.524).

Interestingly, the most prominent formalisation of spatial dimensions, by Henry Poincaré (1912), was made in the context of establishing the most important departure from the Euclidean approach, namely, topology, of which he was one of the founders. It is important to note that his definition of dimension was driven by his pursuit of epistemological questions rather than arising directly from his purely mathematical work (Crilly, 1999, p.12). Poincaré approached the notion of dimension from the perspective of a cut, defining each dimension in terms of the particular spatial configuration of the disruption to the continuity of a particular space. Thus, a line is that space which can be disrupted, or whose continuity can be lost, by a cut at any one point, where a point is a zero-dimensional entity. Likewise, a one-dimensional line cuts a two-dimensional plane and a two-dimensional plane cuts a three-dimensional object. Thus, a space of a lower dimensionality can alter some of the properties of the higher-dimensional space, at the place of encounter. Equally, a space is defined by the dimensions of that which can disrupt its continuity. So higher-dimensional spaces contain lower-dimensional ones, whilst lower-dimensional ones are both included in and can cut higher-dimensional spaces. Most importantly, the space thus defined is continuous, and the coordinates defined by each dimension have no relationship to each other besides defining together the spatial arrangements of a whole (Poincaré, 1912, p.487). It is important to get used to the idea of space in most general terms, away from the physicality of the space we are familiar with, and to think of it as sets of elements with particular properties, or as manifolds.[3] In this sense, a plane is a class composed of all the points uniquely defined by two coordinates. An important clarification about manifolds, as put forward by Kasner and Newman (1940), is that 'a manifold, as usually defined, is stripped of every attribute, except that it is a class' (p.154). In this sense, a three-dimensional manifold is a class of elements 'each of which would require exactly three numbers to identify it – to distinguish it from every other element in the class' (p.154). Dimension is therefore replaced by coordinates, and space becomes the answer to a question of classification.

We can anticipate here the relevance to psychoanalysis of spatial thinking in general and of topology in particular, by remembering that, when we deal with the unconscious, we deal with a set of elements that are structured. In most general terms, a space is a collection of points (elements) that have a particular relation to each other, which is exactly what topology studies, namely, sets (finitely many or countably infinitely many) of 'building stones, assembled together as a structure' (Seifert and Threlfall, 1980, p.5). This is precisely the nature of language, which can also be understood as having a spatial structure.

These notions have been theorised and formalised in various ways (Hurewicz and Wallman, 1948), but what remains as a striking feature is the inductive nature of the ideas at play, as they tried to approximate our intuitive perception of dimension, whereby unidimensional objects (lines) have length, two-dimensional objects have areas, while three-dimensional objects have volume. This particular understanding of space, predicated on measure, is one that is open to most objections, when it comes to thinking spatially about the mind, despite the widespread use of questionnaires and scores in the pursuit of 'measuring' its workings in a number of clinical settings.

As we are going to see in Part III, the very notion of dimension was severely challenged by some mathematicians, in particular by Cantor, who brought about new, more flexible, but also less intuitive, ways of conceptualising space.

Although I will not pursue them further here, it remains important to acknowledge also the existence of fractals (see also Chapter 3), which are a particular type of object whose dimensionality can be a fraction, rather than an integer. Thus, fractals in the plane can have any dimension between zero and two (Falconer, 2013, p.48). For instance, the coastline of Britain has a dimension of approximately 1.21. This is an approximation because the measure is dependent on the scale of measurement. In the body, the respiratory system, blood vessels (and the heartbeat itself) as well as the nervous system are all fractal structures (Falconer, 2013, pp.106–108). The same notion can be applied to the structure of numbers with infinite decimal places (Falconer, 2013, pp.112–115). All this matters because, as it has become more widely accepted since the 1960s, when Benoit Mandelbrot, the mathematician who introduced the term fractal in 1975, noted that irregular objects are to be regarded as the norm rather than the exception (Falconer, 2013, p.120), a notion which is akin to an interest in subjectivity in mathematical terms. A good exploration of the link between fractals and the psyche is offered by Marks-Tarlow (2008).

The most commonly held view, currently, regarding space and dimensions is that we inhabit a three-dimensional physical space, with time as a fourth dimension. For instance, as early as 1901, Hinton pointed out that 'extension on the unknown dimension appears as duration' (p.156), but also posed the question of whether humans are beings with an infinitely minute access to a fourth dimension. He was at that point 'the chief popularizer of hyperspace philosophy' (Blacklock, 2018, p.105), writing as a mathematician at a time when much interest was devoted to notions of higher spatial dimensions not just in mathematics, but also in the more controversial and mystical circles of spiritualism and occult societies which located spirits in the fourth dimension (Blacklock, 2018). One could say that there was a fashion at the time to consider the fourth dimension. Nevertheless, despite its romantic and fantastic appeal, the notion of the fourth dimension is a crucial step in abstract thought. In the words of Kasner and Newman (1940),

'[N]o concept that has ever come out of our heads or pens marked a greater forward step in our thinking, no idea of religion, philosophy, or science broke more sharply with tradition and commonly accepted knowledge, than the idea of a fourth dimension.

(p.131)

Modern physics still grapples with other ways of understanding space, including the view that the world we inhabit might be just a 'shadow' of a more spatially complex world, or that space has many more dimensions – the three 'large ones' we perceive, and one or more other dimensions, small and curled, as posited by string theory (Greene, 2000; Wertheim, 2018). In the words of Stephen Hawking and Leonard Mlodinow, 'we and our four-dimensional world may be shadows on the boundary of a larger, five-dimensional space-time' (Hawking and Mlodinow, 2010, pp.59–60). This hypothesis (called by them the holographic principle) offers a good entry point to the examination of the relationship between spaces of different dimensions, and between body and mind.

4.1 Flatland and beyond

The question of dimensionality is, for me, a very personal one. Years ago, as I was about to begin my first analysis, a friend asked me a simple question: 'Why?' My immediate answer was in terms of dimensions: if I am a two-dimensional being,

I would need to encounter a three-dimensional being in order to learn something about myself. The friend's response was simple as it was mysterious: 'I know just the book for you'.

That book was *Flatland: A Romance of Many Dimensions*, published in 1884, some 20 years before the formulation of the theory of relativity, by Edwin A. Abbott, under the name of A Square.[4] This is the story of a square who calls himself a Square and is a mathematician in his own two-dimensional world. The Square has the brief experience of entering a higher dimension, after which he returns to his own flat world transformed by this adventure, yet unable to convey much about it to his flat people. The book – published in the third dimension of *Spaceland* – begins with a preface to a revised second edition, in which the Square deplores the troubles with higher dimensions and warns his readers that similar struggles also apply to them:

> but in reality you also see (though you do not recognize) a Fourth Dimension, which is not colour nor brightness nor anything of the kind, but a true Dimension, although I cannot point out to you its direction, nor can you possibly measure it.
>
> (Abbott, 2015, p.9)

In the world of *Flatland*, all beings are two-dimensional, and can only perceive each other as a line, as they 'see' each other sideways, that is to say, the inhabitants of this two-dimensional world appear to each other as unidimensional. The story explores in some detail the difficulties that such limited perceptions pose, from efforts to use sight and feeling in order to establish something about the true nature of the other to legal obligations and prohibitions to ensure that order and decency are maintained in Flatland. The story has been primarily recognised as a satire of the society in Abbot's time. However, beyond the impositions and limitations of class and rank that he takes on in this particular way, the question he raises is fundamentally that of the difficulties with perceiving the 'true nature' of a being, if by that we mean a characteristic that exists but does not lend itself to perception in the usual way. Likewise, we are four-dimensional beings that can only perceive, immediately, the three-dimensional presence of others, in their bodies, and struggle with access to the fourth dimension, that of our own minds and of others.

The encounters that the Square has with a higher dimension are akin to a psychoanalytic encounter, whereby something hitherto unknown becomes available and takes the form of irreversible knowledge, but also one which is ultimately impossible to share or pass on to another. The absence of any handbook of psychoanalysis and the ongoing debates around transmission in the profession testify to that very difficulty.

Before his encounter with a being from a higher dimension, the Square has a dream about visiting a world of a lower dimension, *Lineland*, where the inhabitants are lines or dots and can only perceive each other as points, given that they 'look' at each other from only one end. Each being in this world can move along the line, but remains trapped in his/her position relatively to its neighbours. The Square attempts to enlighten the King of this world about the possibilities of movement in a different direction. When he tries to describe a movement sideways rather than along the line, the King is troubled by this idea: 'How can a man's inside "front" in any direction? Or how can a man move in the direction of his inside?' (Abbott, 1995, p.78).

As Matte Blanco points out in his own psychoanalytic reading of this story, the very notions of internal and external change with the dimensions of the space in question (Matte Blanco, 1988, p.303). Yet, with each move to a higher dimension, what is required is a radical shift away from what is known, starting from 'inside', towards a hitherto not accessed direction.

While the Square can comprehend how a point can move through a length to make a line, and how a line can move parallel to itself to make a square, he cannot take this analogy any further until a being from three dimensions, a Sphere, pays him a visit. The Square's initial reaction is no different from that of the King in *Lineland*: disbelief and rage, while proclaiming the knowledge of his world as ultimate. The Sphere takes the Square with him into *Spaceland*, and there the Square finds a way of combining his previous theoretical, mathematical understanding with experience, so that he is soon ready to head for the fourth dimension, at which point it is his three-dimensional guide, the Sphere, who objects that 'There is no such land. The very idea is utterly inconceivable' (Abbott, 1995, p.107), only to concede that anything which seemed to be an encounter with such a higher dimension could only be regarded as 'visions arose from the thought [...] from the perturbed angularity of the Seer' (Abbott, 1995, p.109). The Square remains undeterred in his belief in the possibility of a Cube 'moving in some altogether new direction, but strictly according to Analogy, so as to make every particle of his interior pass through a new kind of Space' (Abbott, 1995, p.109). Analogy is the law that the Square can see at work in the movement from point to line to plane and space, a law that indicates that access from a world in lower dimension into a higher one is through movement in a new direction that is experienced as interior to itself by the being adapted by their senses to its own space of lower dimensionality.

The Sphere also takes the Square to visit *Pointland*, a world of no dimension also known as *Abyss* or *No dimensions*, which, like all the others, is self-sufficient and satisfied with taking its own configuration as a measure of all there is.

After such a tour de force, back in Flatland, the Square is torn between the wish to share his newly acquired knowledge and fear of the authorities determined to silence all talk of higher dimensions as disruptive. He tries to present his learning under the guise of fiction, veils the higher-dimension Space under the name of *Thoughtland*, but ends up in prison nonetheless. His greatest suffering is not so much that particular form of loss of freedom, as it is the impossibility of bringing into his world something that cannot fit in, in any way. Everything in his world is a line, including the paper he writes on. He remains alone with the understanding that there is no way a Cube can be represented as such through just a line. No matter how much he wants to bring the higher-dimensional space into his space of lower

The rigour of spatial dimensions – of shadows and recurrences   69

dimensionality, the Square cannot do so other than by talking or writing about about it, that is to say, he can only accomplish an approximation in language. In this sense, speech and language operate like an additional dimension in relation to those of experience. The prohibition that confronts the Square, according to the rules of Flatland, is merely a veil. Had the authorities allowed him to speak or write freely about spaces of higher dimensionality, the intrinsic nature of his space would have prohibited him from bringing those spaces into Flatland as such. Higher does not fit into lower, but unlike the case of numbers, one does not end up with a fraction of the total, but a set of disparate fragments. There is, in other words, a fundamental asymmetry between spaces of differing dimensionality.

Such is also the nature of the encounter with the unconscious, the importance and the limitation of speech in our clinical work, as well as the structural impossibilities around the transmission of the knowledge at stake. Both the unconscious and languages are structured like spaces, but not in a way that permits any one-to-one mapping between them.

A common feature of the Square's encounters with all the beings inhabiting worlds of different dimensionality compared to his own is that something is experienced as partial and self-contradictory, and deemed impossible according to the goings-on in the lower dimension which is hosting the encounter. For instance, when the Sphere first comes to Flatland, it appears to the Square as a circle of varying size, arriving from nowhere (i.e. from above). This is impossible by local standards, no Flatland being could be all of those circles at once, but in the third dimensions all the circles are parts of one single entity, namely, the Sphere. Given the limitations of living in a two-dimensional spaces, all the Square can perceive is a sequence of lines of differing curvatures, which he interprets to be a sequence of circles of varying dimensions. The illustration that Abbott himself offers shows the arrival of the Sphere in *Lineland*, where it becomes a segment of varying length as it 'travels' through, all of which from within the line can only to be perceived as a single point (as seen from one end of the segment or the other) – see Figure 4.1
In this charming story, as well as in the mathematics that underpins it, movement between spaces of different dimensionality is entirely conceived of in terms of dimensional analogy: the limitations of a line in the plane translate to that of a

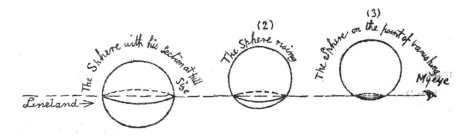

Figure 4.1 Sphere passing through linear space.

two-dimensional shape in three dimensions, and so on. As Blacklock (2018) shows, the text uses analogy and comments on it at the same time (p.75). Of course, analogy is not free from pitfalls, given that it relies on a principle of continuity founded on the idea of uniformity in nature (Blacklock, 2018, p.84).

Yet the analogy can be taken one step further, to consider how the encounter with an entity from the fourth dimension would be experienced in our usual three-dimensional space, something that the Square himself was ready to consider, but unable to conceive of. This is precisely the setup of another mathematical story that links dimensions to the mind: '*It came from... higher space!*', a 1960s retro format comic by Allan Moore, published in the early 1990s in a comic collection with a title of Freudian resonance: *Tales of the Uncanny*. This is the story of a superhero known as the Hypernaut, who encounters a threatening being from the fourth dimension (Moore, 1993). The Hypernaut is a brain connected to a disposable body, and lives on a space station in the shape of a Penrose triangle (Penrose and Penrose, 1958), a geometrical construction as depicted in Figure 4.2, and described as 'impossibility in its purest form' (Rooney, 2013, p.118), hovering on the dark side of the moon. This shape is one where up and down are not distinguishable, and neither are an inside and an outside.

On his way to finding the source of the disruption caused by the unexpected visitor to his spatially complex home, the Hypernaut checks the 'verizontal' and 'inside-outerior' of his base – terms with much Lacanian resonance, in the way they condense the vertical with the horizontal, or the inside with the outside. The aggressor from the fourth dimension can only be perceived in disconnected sections that approach from more than one direction at the same time, and the Hypernaut struggles to grasp the logic behind these apparitions. It is only as he gets thrown by the intruder through a section of Flatland, which is hanging as a decoration on the wall of his Penrose base, that the Hypernaut comprehends what is at stake. He then works out a way to conquer the invader: by filling its mind with three-dimensional, conscious facts which, it turns out, the monster finds

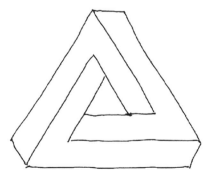

Figure 4.2 Penrose triangle.

unbearable. In this way, the conscious mind manages to send back the threatening, disconnected and powerful (unconscious) manifestations to where they came from. In this case, the solution is that of repression, back into the fourth dimension.

In a modern sequel to *Flatland*, titled *Flatterland: Like Flatland, Only More So*, Ian Stewart (2002) takes the exploration of dimensions even further, with much humour built on rich linguistic fluidity, while consistently pertinent and accurate. He explains the concept of dimension as follows: '"Dimension" is a geometric way of referring to a variable. Time is a non-spatial variable, so it provides *a* fourth dimension, but the same goes for temperature, wind-speed or the number of termites in Tangentia' (Stewart, 2002, pp.46–47). In this respect, a message is 'a point in multidimensional space' (p.53), while 'appearances can be deceptive – or rather [...] absolutely correct, but irrelevant' (p.62). The important thing is not to focus on complex behaviour, but on the simple rules that govern it, as in the case of fractals. After a tumultuous adventure across geometries and spaces, in the company and under the guidance of the Space Hopper, the main character, a young line from *Flatterland*, returns home transformed by the experience: 'Never again, though, would she imagine that just because Flatterland *looked* like a plane, it must therefore *be* a plane' (2002, p.288). While exploring distant and complicated worlds and topological spaces, she comes to learn something about her own world: although in her homeland she is only a line, that line perceived there is only the side of a shadow pentagon, the rest of which could not enter her two-dimensional world, but merely remained external to it. This is not a Jungian shadow of a darker aspect of the same line; neither is this a shape in an extra spatial dimension. Rather, the rest of its being is in some kind of supersymmetry realm, one that cannot be visited, but only inferred. The search for the hypothesised force-carrying Boson particle at Cerne is a search for something of this kind (e.g. http://home.cern/about/physics/supersymmetry).

The unconscious, imaginary numbers, the fourth dimension, the Boson particle are all unifying hypotheses that can be inferred on the basis of incomplete observations. Nevertheless, each of these presuppositions in its own way allows deepening an understanding that would otherwise not be available.

4.2 From higher to lower dimensions: projection

We regularly speak of the world that we live in as three-dimensional, yet the eye perceives this through a two-dimensional retina.[5] Some see mathematics as an extension to the limited perception that vision alone affords us (Cohen-Tannoudji and Noël, 2003, p.130). In other words, we are not biologically equipped to perceive space other than as a projection, but we can make up for this deficiency by using the rigour of mathematical thinking. Or, in Lacanian mode, the Symbolic is there to address that in which the Imaginary fails.

A common way of thinking about the interplay between spatial dimensions is through the notion of projection, that is to say, through considering the shadow cast by the object situated in one space onto a space of lower dimensionality. Such a translation from one space to the other is, at best, reductive, and often impossible. The impossibility of bringing a four-dimensional object on to the page as a drawing is such an instance, which precludes the use of the Imaginary as a principal avenue for engaging with this possibility of such a space, and invites rigorous symbolic effort before it yields any understanding. While the two-dimensional image of a three-dimensional shadow of a four-dimensional object is possible to create, the very spatiality at stake is lost in such a representation. This is precisely the constraint faced by the Square in his writing.

The fact that a three-dimensional object casts a two-dimensional shadow on a plane was used by Plato in his allegory of the cave, in book VII of the dialogue on the Republic (Plato, 2015). In the scenario he describes, slaves bound to face the back wall of a cave remain incapable of understanding the world they inhabit as anything other than the collection of shadows projected in front of them. Yet, behind the slaves there is a space where three-dimensional copies of three-dimensional objects that exist outside the cave are paraded in front of a light, so as to cast their limiting shadows for the slaves. Once freed from their bounds, the slave can not only see the puppetry at play, but also, emerging from the cave, they can encounter the very objects whose copies were used to cast the shadows, as well as the light that illuminates this entire succession of representations. In his exploration of freedom and the place of knowledge, Plato constructed a hierarchy between the limited perceiving and naming of a projected shadow of a copy of a living thing, the encounter with the 'puppet copy' of the real thing, the real thing in the light outside the cave, and the very light that illuminates it. Thus, since Plato, we are easily reminded that inhabiting a world of shadows makes one a prisoner destined to face away from the source of knowledge itself.

Part of this limiting outcome is linked to the spatial position of the subject in relationship to that which is to be perceived and known in some way. The prisoners in Plato's allegory begin life tied into positions that stops them from seeing anything other than the shadows projected in front of them. Once freed from their bounds, they can move around and encounter the world that surrounds them in new ways. The removal of this gives them access to an additional degree of freedom in the mathematical sense. In Plato's allegory, the real world is three-dimensional, while the perception of the slaves is limited to two-dimensional shadows. Likewise, mathematically, '[A]s the real world is to this shadow world, so is the higher world to our world' (Hinton, 1901, p.2).

The work of Ludwig Schläfli, another contemporary of Freud's, on the four-dimensional equivalents of three-dimensional polyhedra has been brought to life through recent developments of computer-generated representations (e.g. http://www.dimensions-math.org/). What is striking about the three-dimensional shadows of four-dimensional objects, shadows which are themselves three-dimensional objects in their own right, is how the relative positions of the components change: in front and behind, above and below, inside and out, are interchangeable, and constantly flow into each other. These unstable position coordinates are not in the nature of the four-dimensional object, but simply a consequence of the limitations in the perception of such objects, with a two-dimensional visual apparatus, of limiting representations in three-dimensional space. This is, again, the fundamental asymmetry between spaces of unequal dimensionality.

Shadows or projections create distortions and overlaps, a loss of information that warps the perception of the higher-dimensional object. In a world of shadows, any perception is a misperception.

4.3 From lower to higher dimensions: repetition

Space of higher dimensionality is generated by movement of a lower-dimensional space in a direction not contained within itself. Thus, the movement of a point away from itself creates a line, in the way that dragging the point of a pencil on paper always does. Drawing is an act of accessing spaces of a higher dimensionality, in the movement it creates from the 'nothing' of the dimensionless point, into the line and beyond. The sliding of a line away from itself produces a plane, whilst the movement of a plane in a direction it does not contain creates three-dimensional solid space. In Hinton's definition, 'space is that which limits two portions of higher space from each other' (Hinton, 1901, p.4). By induction, he proposes that 'our space will generate the higher space by moving in a dimension not contained in itself' (Hinton, 1901, p.4). In other words, the world we live in can be understood as a three-dimensional 'edge' of a four-dimensional space.

Mathematically, movement from a space of a lower dimensionality to one of a higher dimensionality does not seem to have its own name as an operation, although it can be thought about as some kind of expansion. Up to the third dimension, this is something that occurs through the introduction of a new dimension positioned orthogonally to the pre-existing ones. Thus, as we have seen in Chapter 3, in the case of complex numbers, the addition of a vertical line to a pre-existing horizontal number line turns the space from a line to a plane in which movements outside the space of the line of real numbers become possible. The addition of a third dimension, orthogonal to the plane created by the two axes, introduces the possibility of movement upwards and away from the plane of the page on which the two axes are drawn. The analogy can be carried further only intellectually, as the physicality of the space we inhabit only allows for time as a fourth dimension. In the way that a two-dimensional being could not escape the confines of the boundary of a circle drawn around them, without breaking it, in three dimensions a being trapped inside a sphere could not escape it without cutting it either. Yet stepping outside it along a fourth dimension would not pose any constraints, in the same way that knots in three dimensions can be unknotted in four dimensions without cutting. Any such movement is one whereby the 'interior' of the entity in question moves away from itself, an operation akin to that of the work of analysis.

While it is not possible to access fully a higher dimension from a lower one, it is possible to experience traces of movement in the higher dimension as a succession of repeating instances. Indeed, what can appear in three dimensions as a succession of similar but altering objects can be understood as the movement of a single entity located in four-dimensional space and which cannot be contained in three dimensions in its entirety. As Hinton expressed it: 'life, and the processes by which we think and feel, must be attributed to that region of magnitude in which four-dimensional movements take place' (Hinton, 1901, p.19). Although he laments the difficulty with showing life to be a phenomenon of motion, Hinton does assume 'the human soul' to be 'a four-dimensional being' 'capable in itself of four-dimensional movements' (p.20). In his view, a being with such attributes 'would have a consciousness of motion which is not as the motion he can see with the eyes of the body' (Hinton, 1901, p.34). Hinton follows the argument that Aristotle's distinguishing between matter and form implies the existence of the fourth dimension (p.37), and pursues this notion mathematically through the work of Lobachevsky and Bolyai, the founders of (post-Euclidean) hyperbolic geometry, who distinguished between the laws of space and the laws of matter (p.55). Like they did in relation to the mathematics that came before them, and as Lacan often insisted to be necessary in psychoanalysis, Hinton promotes a 'perfect cutting loose' (p.57) from familiar intuitions and from sense, and draws attention to the fact that 'life is essentially a phenomenon of surface' (p.74), as '[O]ur three-dimensional world is superficial' (p.84). This was, as we have seen in Chapter 2, Freud's take on psychic phenomena, and also resonates with Ragland's (2002) reading of Lacan as affirming that 'the foundation of the surface is at the base of everything we call the organisation of the form' (p.122).

Yet to many, including mathematicians, formal spatial formulations outside the norms of Euclidean thought remain a kind of 'fairyland of geometry' (Newcomb, 1897). Given that there is no scientific proof that space is three-dimensional, Kasner and Newman (1940) argue that the notion of four-dimensional geometry is something that needs to be regarded as a game, like chess, where one is not concerned with plausibility and its testing against reality, but with its internal consistency (p.117).

The representation of higher-dimensional objects in lower dimensions has two significant consequences, both with complex clinical equivalents and implications. First, *overlaps* emerge. The simplest way to recognise this is to consider a most common occurrence, where the representation is a projection, namely, the shadow that one casts. From a three-dimensional body, all that remains is a two-dimensional stain, where no distinctive features can be identified, and where something present in three dimensions can appear to be missing in two dimensions, for instance, an arm on which the light does not fall, because it is on the other side of the body. Second, *repetitions* occur. For instance, a circle, which is a two-dimensional object, can be represented as a line in one-dimensional space, as shown in Figure 4.3.

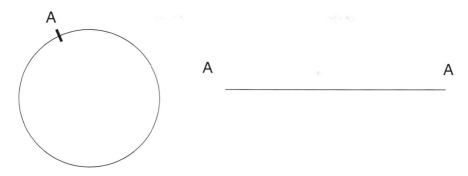

Figure 4.3 Two-dimensional and unidimensional circle representation.

A single point A on the circle has two occurrences on the line representation, which is the equivalent of cutting the circle at point A and straightening it out, while specifying that the two ends are identified. Equally, the line representation is unidimensional and can only be turned into the familiar representation of a two-dimensional circle once the vertical dimension is added and the ends of the segment can be curved so that A becomes a single point, again. What appears in two dimensions to be a repetition of A is in two dimensions a single instance of a different whole.

In both Freudian and Lacanian terms, repetition is the manifestation of the unconscious, present in every subject. Its universality is indicative of a structural underpinning. The inevitability thus at play is often seen as an automatism that evades the intention of the subject. The main implication of regarding this as a necessary consequence of spatial structure is that repetition is no longer to be regarded as a symptom, but a consequence of bringing into experience a sequence of traces pertaining to one single, unitary inscription lodged in a space of higher dimensionality.

In his later work, Lacan distinguishes between repetition produced by the nature of the Symbolic through the operation of signifiers and repetition rooted in the Real, in that which escapes symbolisation (Lacan, 1979). In this sense, it is the latter category that reveals the inherent spatiality of the unconscious in relation to that of the space of physical lived experience, whilst also recognising language as structurally spatial. That is to say, repetition is structural.

4.4 Spatial dimensions in psychoanalysis

Direct references to spatial dimensions in the psychoanalytic literature are sparse, with only Mate Blanco and Lacan as contributors to a modest offering on this theme. Although Matte Blanco and Lacan were contemporaries, they seem to have remained unaware of each other's work, so no dialogue was, unfortunately, possible between them.

Chronologically, Lacan approaches this question first, but Matte Blanco is the one with the more extensive take on the matter.

Thus, Lacan takes up the question of spatial dimensions in the early 1960s, in his seminar on *Identification*. Here, he acknowledges the limitations of his own contributions in appealing to topology: 'I could not really give you anything other than some sort of projections, in the way one tries to show in another space four-dimensional figures which cannot be represented' (p.345).

More specifically, in the session of 2 May 1962, he makes clear his view that, in psychic terms, we only have access to two dimensions (Lacan, 2020, pp.322–323). This is completely consistent with Freud's view of the psyche as a surface which we have examined in Chapter 2. At the same time, Lacan raises the question of embeddedness, as he stresses that the psychic surface is immersed in a space which is not at all like the one we would imagine on the basis of the visual experience of the specular image (Lacan, 2020, p.323). Nevertheless, Lacan deplores the casual use of references to the fourth dimension as a link to science fiction and its bad reputation. It is not clear what he is referring to, here, in terms of psychoanalytic relevance.

Matte Blanco links mathematical notions of dimensionality to the very essence of the unconscious. In a book published in 1988, some 13 years after the publication of his tome on the particular logic of the unconscious (*The unconscious as infinite sets*), in a section titled '*Why is the Freudian unconscious unconscious?*', he poses the question directly:

> 'could it not be that the unconscious deals with a number of dimensions greater than that with which our conscious thinking can deal? And could it not be that, if we were able to think in terms of more dimensions, then all the strange behaviour of the unconscious could easily enter into our consciousness and have a logic of its own?
>
> (Matte Blanco, 1988, p.90)

Indeed, it is precisely to the examination of this possibility that the present chapter and the next are devoted to.

While Matte Blanco is very detailed and explicit in conveying his reasoning, Lacan is characteristically obscure and often makes use of mathematical formulations of space, without explicitly acknowledging that, as we will see in the next chapter and beyond.

The notion of four- (and higher-) dimensional space poses the question of whether such space is in the realm of physical reality, or just a theoretical construct. Blacklock (2018) offers a pertinent response:

> According to Kant, space is a priori because it comes before the empirical, it is not sensed. Higher-dimensional space, while it is not sensed, does not come to us in the same way. It is a product of the understanding alone, an a posteriori concept

without empirical content. We may concede that it is possible to imagine higher space but there remain significant obstacles to representing what has never been sensed.

(p.168)

This all holds as long as we do not include dreams under the heading of sensing.

In the next chapter we turn to a closer examination of how the notions of spatial dimension, projection and expansion introduced in this chapter can be employed to understand the nature of the unconscious and of the workings of psychoanalysis. A consideration of the limitations of this approach leads, in Part III, to the exploration of an alternative mode of understanding space mathematically, and of its psychoanalytic relevance, as we turn to an exploration of topology.

Notes

1 Against this backdrop, theories rather than just statements about dimensions did not emerge until the late 1970s (Crilly, 1999, p.3).
2 Blacklock (2018) identifies Hinton as 'arguably the least well-known yet most influential theorist of higher space of the late nineteenth century' (p.10). Both authors use the notion of higher space as a contraction for higher-dimensional space.
3 All the manifolds considered here will be two-manifolds, that is to say, surface-like objects such that each and every of their points has a disk-like neighbourhood (Adams, 2000, p.73).
4 His middle name was also Abbott, so Abbott Abbott, AA, mathematically A^2, which is read A squared: A Square.
5 Hawking and Mlodinow (2010) explain succinctly the limited nature of this:

> There is a blind spot where the optic nerve attaches to the retina, and the only part of your field of vision with good resolution is a narrow area of about 1 degree of visual angle around the retina's centre, an area the width of your thumb when held at arm's length. […] the human brain processes that data, combining the input from both eyes, filling in the gaps on the assumption that the visual properties of neighbouring locations are similar and interpolating.
>
> (pp.62–63)

References

Abbott, E.A. (1995[1884]). *Flatland: A romance of many dimensions, by A Square*. CreateSpace Independent Publishing Platform.

Adams, C.C. (2000). *The knot book: An elementary introduction to the mathematical theory of knots*. Providence, RI: American Mathematical Society.

Blacklock, M. (2018). *The emergence of the fourth dimension: Higher spatial thinking in the fin de siècle*. Oxford: Oxford University Press.

Cohen-Tannoudji, G. and Noël, È. (eds.) (2003). *Le réel et ses dimensions*. Editeur: EDP Sciences.

Crilly, T. (1999). The emergence of topological dimension theory. In James, I.M. (ed.). *History of topology*. Oxford: Elsevier, pp.1–24.

Falconer, K. (2013). *Fractals: A very short introduction*. Oxford: Oxford University Press.

Greene, B. (2000). *The elegant universe: Superstrings, hidden dimensions and the quest for the ultimate theory.* London: Vintage Books.
Hawking, S. and Mlodinow, L. (2010). *The grand design.* London: Bantam Books.
Hinton, C. (1901). *The fourth dimension.* Swan New York: Sonnenschein & Co. [reprinted by Forgotten Books in 2002].
Hurewicz, W. and Wallman, H. (1948). *Dimension theory* (PMS-4). Princeton, NJ: Princeton University Press.
Kasner, E. and Newman, J. (1940). *Mathematics and the imagination.* New York: Dover Publications.
Lacan, J. (1979[1973]). *The four fundamental concepts of psychoanalysis.* London: Penguin Books.
Lacan, J. (2020). *L'identification: Séminaire 1961-1962.* Éditions de l'Association Lacanienne Internationale. Publication hors commerce.
Marks-Tarlow, T. (2008). *Psyche's veil: Psychotherapy, fractals and complexity.* London: Routledge.
Matte Blanco, I. (1988). *Thinking, feeling, and being: Clinical reflections on the fundamental antinomy of human beings and world.* London: Routledge.
Moore, A. (1993). It came from a higher dimension. *Tales of the Uncanny* 1963, Book Three. Canada.
Newcomb, S. (1897). *The fairyland of geometry and other essays in science,* ed. David Stover (ed.). Rock's Mills Press.
Penrose, L.S. and Penrose, R. (1958). Impossible objects: A special type of visual illusion. *British Journal of Psychology*, 49, pp.31–33.
Plato (2015). *Great dialogues of Plato,* trans. W.H.D.Rouse. New York: Signet Classics.
Poincaré, H. (1912). Pourqui l'espace a trois dimensions. *Revue de métaphysique et de morale*, 20(4), pp.483–504.
Ragland, E. (2002). The topological dimension of Lacanian optics. *Analysis*, 11, pp.115–126.
Rooney, A. (2013). *The history of mathematics.* Arcturus Publishing Ltd.
Seifert, H. and Threlfall, W. (1980). *A textbook in topology.* New York: Academic Press.
Smith, D.E. (ed.). (1959). *A source book in mathematics.* New York: Dover Publications.
Stewart, I. (2002). F*latterland: Like Flatland, only more so.* New York: Basic Books.
Wertheim, M. (2018). *Radical dimensions.* Aeon Essays. https://aeon.co/essays/how-many-dimensions-are-there-and-what-do-they-do-to-reality (Accessed: January 2023).
* Dimensions. http://www.dimensions-math.org/ (Accessed: 8 December 2022).
* http://home.cern/about/physics/supersymmetry (Accessed: 20 January 2023).

Chapter 5

The unconscious as inaccessible and the exclusivity of the fourth dimension

The world of modern physics is one of gaps and discontinuities, 'where space is granular, time does not exist, and things are nowhere' (Rovelli, 2014, p.77). It is difficult to distinguish this from a description of dreams. Indeed, the same author, a physicist, makes this very link in a later work, where he reflects on the quantum understanding of the world as one that presents us with '[A] reality which seems to be made of the same stuff our dreams are made of' (2017, p.73). Intuitively, existence is linked to location: 'Many statements of existence work in such a way that they define location' (Gabriel, 2016, p.7). On the contrary, quantum physics – which is far from intuitive – posits that the particles that make up the world 'do not have a pebble-like reality' (Rovelli, 2014, p.30). They are nothing more than 'elementary excitations of a moving substratum' and 'everything that exists is never stable, and is nothing but a jump from one interaction to another' (*ibid*). Something in this resonates with Freud's (1950) model of the libido in his early *Project for a Scientific Psychology* (SE1), where movement of excitations and problems with locating the processes at play were the focus of his attention.

The unconscious, like the mind, is difficult to place spatially. While neuroscience has located the mind firmly in the spatial domain of the brain, psychoanalysis has moved away from the question of the anatomical location for mental processes, by positing the unconscious more as a function than as a place where things happen. Without entering into any philosophical debates concerning ontology, we can at least call upon one relevant approach, and consider that, in the light of thinking from New Realism, 'perspectives onto things are features of the things themselves' (Gabriel, 2016, p.9). If the unconscious is a 'thing', then its lack of spatiality in the ordinary sense can be seen as one of its features. But what of its trans-spatiality, its features as a structure 'located' in a fourth dimension? What can an enhanced conception of space contribute to our understanding of what we experience as inaccessibility?

In Tibetan Buddhism, the mind is spatial, with a sky-like nature, penetrating in all directions and pervading everywhere (Sogyal, 2002, p.157). In this philosophy and belief system, the mind is space itself, in that it generates the body and contains both life and death: 'life and death are in the mind, and nowhere else' (Sogyal, 2002, p.47). This implies that, as a space, the mind has an additional dimension in relation to the perceived spatial world, in order to be able to both accommodate it and surpass it by generating thought about it. In the words of one of my analysands, 'the mind is different, but we don't know how, no one has been outside of it to see what is there'. The extra dimension of the mind offers a particular kind of '"outside'. In the same context, knowledge is also a spatial experience, in that the removal of ignorance leads to the 'wisdom of all encompassing space' (Sogyal, 2002, p.284). In this sense, we do not *have* a mind, as much as we *are* in one.

So what could be the relationship between the mind in Tibetan Buddhism, the unconscious according to Freud and four-dimensional space? In the same way that fragmented conscious experiences can appear as disconnected and unintelligible when considered one by one, in isolation, but unified and decipherable once the hypothesis of the unconscious has been made (Freud, 1915, p.167), so the recognition of the fourth dimension can inform about the coherence and totality of apparently fragmented partial encounters in three dimensions. In other words, the experience of three-dimensional space is immersed in the four-dimensional space of the unconscious, which in turn 'sits' in the higher-dimensional space of pure consciousness, as envisaged by many spiritual traditions.

In this sense, we can think of the unconscious as something that, at the level of experience, can only be perceived in fragments, always discontinuous, incomplete and elsewhere. This apparent discontinuity and inaccessibility is the effect of an impossibility intrinsic to engaging with a four-dimensional space from three-dimensional perception limited by the structure of the space inhabited by the body. The lack of one-to-one mapping between the unconscious and the body is also a factor in the 'translation' from body to experience and to language, whose spatial structure also intrinsically excludes completeness and continuity.

What is traditionally regarded as the infinite nature of the unconscious could therefore be thought about in a different way: rather than a domain akin to an interminable sequence of elements (à la *analysis terminable and interminable*), the unconscious can be conceptualised as a domain whose totality can never be apprehended not because its vast size, not because there is not enough time to do so, but because its totality resides in another dimension and is therefore structurally out of reach. Indeed, this is consistent with Lacan locating the field of the Other in an 'indeterminate place' (Charraud, 1997, p.46). To be clear, four-dimensional space can be conceptualised either as finite or as infinite, so this is not a question of shifting infinity from one dimension to the next. What the concept of higher-dimensional space can do is offer another way to think about indeterminacy in the context of clinical work. Whether finite or infinite, four-dimensional space is never reducible to three-dimensional space. This is precisely recognised by Freud in his notion of the navel of the dream, as we have seen in Chapter 3.

This consideration alters the nature of the inaccessibility: it is not merely a matter of inconvenience in terms of distance, but a structural otherness of a dimension which is by definition external to and radically different from the three familiar ones. We could call this a dimensional inaccessibility. In formal terms, the one thing that the axes of coordinates have in common is one point, the so-called origin, their point of intersection, conventionally located at zero, that is to say, nowhere in particular. This implies that access to this new dimension can be gained from anywhere, which is precisely how psychoanalysis works: the patient comes and speaks, and that offers a way in, no matter what they say, no matter what they start with. For Lacan (1979), zero has 'the character of an absolute point with no knowledge' (p.253).

As Caussanel (2011) clarifies, in any representation of space, several spaces and sets of dimensions are at stake at the same time: the drawing of a cube on a page is a two-dimensional representation on a two-dimensional support, which is immersed in a three-dimensional space (where the book to which the page belongs is). The space of representation is different from the space of imagination, where we 'see' the flat lines turning into the contours of a three-dimensional object, which is itself different from the space of the object thus represented. In Lacanian terms, only when we introduce the space of representation (the Symbolic) do we recognise that what we took for the Real was only the Imaginary (Caussanel, 2011, p.89).

Abbott's *Flatland* clearly depicted (see Chapter 4) how the dimensionality of each space imposes restrictions on what a being inhabiting it can perceive. The three-dimensional Sphere that visited the world of the Square could apprehend his flat world in its totality, in the same way that we can see at once an entire floor plan drawn on a page.

The human eye is two-dimensional not as organ, but in that it produces two-dimensional images from which the brain extrapolates the third dimension of depth. A three-dimensional eye of a four-dimensional being could see all the sides of an opaque box at once, as well as what is inside the box. This is particularly relevant, given the idea of inside-outside that pervades much of psychoanalytic thinking, in particular within the IPA tradition, which distinguishes between an internal and an external world of the individual, each populated by their own specific objects. Writing from within this tradition, Matte Blanco emphasises that 'the concepts of internal and external must be defined in relation to the dimensions of the space' (Matte Blanco, 1988, p.303). Freud himself recognised two realities – an internal and an external one and, following Kant, concluded that the only one that stood a chance of being known was the internal one (Freud, 1915). Since Freud, the means by which this internal reality could be apprehended was psychoanalysis itself (Nasio, 2010, p.11), which offered the means to engage with the ways in which each subject approaches this reality, namely, through symptoms and phantasy (p.13). Analysis offers the best approximation to stepping outside the space it examines from 'within', merely by centring on the awareness of the nature of the impossibility of what it tries to accomplish.

Any particular conception of space at play has a lot to do with the ways in which the relationship between the inside and outside is understood and between elements that can appear in paradoxical positions, spatially or temporally. Whilst most post-Freudians retained a simplified formulation of space, where one can speak of a container and a contained (Bion), an external and an internal object (Melanie Klein), Lacan challenged at once both such a conception of space and the psychoanalytic implications derived from it. Thus, Lacan's formalisation of the unity between inside and outside through the Möbius strip is the closest we can get to understanding something that we are not biologically equipped to see. The limited perception predicated on notions of 'internal' and 'external' has a profound impact at the level of being, where an insoluble sense of discontinuity persists. In this sense, Lacan's mirror stage model (Lacan, 1949), in its formalisation of the encounter of the baby with their own image as reflection, one of his most widely known contributions, is even more relevant as 'a fundamental aspect of the structure of subjectivity' (Evans, 1996, p.115). In his words, the two-fold value of this model is in that it 'marks a decisive turning-point in the mental development of the child', as well as typifying 'the essential libidinal relationship with the body-image' (Lacan, 1951, p.14). It is not just that the baby, developmentally, makes use of an image (external) to obtain coherence (internal), but that humankind is perpetually under-equipped to perceive the world it inhabits other than in a simplified way. Perception alone, or the realm of the Imaginary, is not sufficient. It is the Symbolic, of which mathematics is one of the purest manifestations, that enables us to access our internal externality, our inner beyond, or, to use Lacan's term, our extimacy (*extimité*, see Evans, 1996, pp.58–59).

Dreams bring, at the level of experience, a very different kind of engagement with space. For the dreamer, any sense of discontinuity manifests not so much in the dream itself, as it does at the point at which they try to put it into words, when the logic of the dream-space fails to fit in with the ordinary notion of 'making sense'. As one patient commented, the language of dreams is something that part of us understands. Our waking sense of space differs radically from that which we encounter in dreams. One could argue that so many dreams are 'forgotten' because their structural four-dimensionality cannot always be collapsed into a three-dimensional representation which can then be put into words, and incompletely at that. That is to say, discontinuity appears as a consequence of the move to a lower spatial dimensionality, rather than it being intrinsic to the dream world itself. In the recounting of dreams this is often captured by the dreamer occupying multiple positions, as participant and observer at the same time, as subject and as object.

The spatial quality of condensation in dreams is a familiar one. Freud (1930) gives a detailed description of the mind as multidimensional in *Civilisation and Its Discontents*, where he likens it to a version of Rome where the architectural present and past occupy the same place, with multiple buildings in the same location, and where time is replaced by spatial coexistence (pp.70–71). Indeed, the bodily perception of space and of the use of the body as reference point for it, a bit like the origin for a set of axes of coordinates, is different in dreams: the dreamer can be in and outside the dream scene at the same time. People and objects can occupy the same space in a way that they cannot in three-dimensional space, with elements that overlap, whilst remaining distinct at the same time. When things appear to be mutually inside each other, as they do in dreams (e.g. Matte Blanco, 1975), their actual spatiality can be understood in terms of dimensions. The same can be said about overlapping images, a common feature of dreams explored in great detail by Freud (1900[1]).

All these 'errors' are defining of the 'forcing' of four-dimensional entities into three-dimensional space. In this sense, the exclusivity of the fourth dimension refers to what structurally remains out of the reach of perception, which remains discontinuous, only to become fragmented further, once translated into language.

After a brief introduction to the notion of four-dimensional space mathematically, we turn to an exploration of the relevance of such a structure to the unconscious and therefore to the clinical process concerned in the work of analysis. Finally, in preparation for the topological exploration of space in Part III, we conclude by considering the most important challenge posed by the Symbolic to the very notion of dimensionality, by examining Cantor's fundamental questioning of the difference between one and two dimensions, as he established an astonishing symbolic equivalence between the two.

5.1 Hyperspace – a primer

Euclidean geometry is a world populated by points, lines, planes and volumes. Although these concepts have numerous applications in reality, for instance, in engineering and architecture, it pays to remember that all the entities upon which geometry operates are merely presupposed abstract conventions. Once they are accepted as given, elaborate constructions can emerge and be examined. It is interesting to note that the entire reasoning about dimensions rests upon an intuitive understanding of these key geometric concepts, whilst all of them remain basically undefined terms in Euclidean geometry (Downing, 2009).

The most common understanding of Euclidean space is as an aggregate of points. A point is a particular location in space, devoid of any dimensions, as by convention it has no height, width or thickness. Such an aggregate of non-dimensional points can be one-, two- or three-dimensional according to how many numbers are necessary to determine or locate any of its elements (Fitch, 1909a, p.43). Thus, a point has no dimensions. A point in space is somewhere, but not something. A line has one dimension, given that one coordinate is enough to locate any one point on it – think back to the edge of a ruler we considered in Chapter 3. Planes are two-dimensional (latitude and longitude are sufficient to locate any point on a map), while volumes span three dimensions (adding elevation, for instance, in a relief map). In this sense, dimension is understood as extension in space.

As Cutler (1909) rightly reflects, the notion of dimension is easier to describe than to define. He does that most succinctly when he expresses the idea of three-dimensional space thus: 'starting from any point in space, we may reach any other point by proceeding successively in three directions at right angles with one another' (Cutler, 1909, p.62).

Departures from Euclidean geometry abounded in the 18th century, stemming primarily from dissatisfaction with Euclid's axioms about parallel lines, which prompted mathematicians to consider worlds defined not simply by straight lines but by curves with particular properties (e.g. hyperbolic or elliptic geometry), as well as higher dimensionality which, in turn, can encompass both Euclidean and curved geometries. Both the established Euclidean and these relatively new alternative geometries referred to spaces that could be defined as unbounded continuums of geometric entities (Fitch, 1909b, p.55). As Riemann's work showed, the way we think about space can be radically altered if we avoid confusing infinity with unboundedness (Kasner and Newman, 1940, p.148). In work that paved the way for Einstein's theory of relativity, the young German mathematician of the 1800s insisted that 'unlimitedness must be distinguished from infiniteness' (Riemann, 1854, p.423). In terms of the implications from this for analysis, the distinction lays the ground for the difference between inaccessibility and impossibility, which we are going to consider further in Section 5.3, and then in Chapter 6. Suffice it to say that the way we think about space has an impact on the way we approach the unconscious and, therefore, our clinical work, whether we are aware of that or not. This is an invitation to engage explicitly with the spatiality of our work.

The main question of interest here is that of the relationship and interplay between spaces of varied dimensionality. The transition from a lower dimension to a higher one can be understood as movement in a direction radically away from the lower dimension. Thus, as we have already seen, a point that moves away from itself creates a line – think of the tip of a pen as the point and then of moving it away from its initial position along the surface of a page. A line that moves parallel to itself creates a plane (think of the edge of an ice-scraper moving parallel to itself, leaving behind a clear surface on a frosted windscreen). Finally, a plane moving outside itself produces a volume, which is the mathematical, purely abstract, counterpart of flat-packed furniture.[2] Although physically impossible to represent, construct or even to imagine, four-dimensional space is conceived of as being generated by the movement of a volume outside of itself, in a direction it does not contain, to create a hyperspace. Given that '[A]ny space is that which forms the boundary between two portions of a higher space' (Fitch, 1909a, p.44), the three-dimensional space we physically inhabit amounts to a boundary of four-dimensional space, or hyperspace. Equally, our three-dimensional space is a hyper-surface, a surface with an added external dimension. Although not arranged as simply or as neatly as a set of Russian dolls, 'Lower spaces are contained in higher' (Fitch, 1909a, p.51). The notion of four-dimensional space is, in this sense, nothing more than a generalisation of three-dimensional space to a higher number of parameters.[3] Such a space can, by definition, contain an unlimited number of three-dimensional spaces (Chambers, 1909, p.204), one of which could be our 'ordinary space' (p.206).

As the introduction to a collection of mathematical essays on the theme of four-dimensional space, published in 1909 by the *Scientific American*, concludes, the existence of such a space cannot be either proved or disproved (Manning, 1909, p.40).

Yet, like the hypothesis of the unconscious, it is one well worth exploring in terms of its properties and of the implications it can yield.

Furthermore, as the winner of this essay competition clarifies: 'Real physical space cannot be said to be either Euclidean or non-Euclidean. Geometry therefore throws no light on the nature of real space' and remains 'a construction of pure thought, a branch of pure mathematics' which, in turn, 'is concerned with implications, not applications.[...] As applied, geometry, in short, is not certain, but useful' (Fitch, 1909b, p.58). It is precisely on that basis that this book proposes the consideration of the unconscious as space.

It is important to keep in mind that there is always a direct correspondence between geometry and algebra, in which the visual qualities of geometry can make many notions appear less abstract. In other words, the symbolic construction of the algebraic expression has a visual counterpart that can often facilitate our grasp of what is at stake. This is not the case in higher dimensions, where only algebra can reach, without the visual being able to follow. Indeed, Camp (1905) proposes that the fourth dimension is a purely mathematical concept (p.118), with the general acceptance of three-dimensional space as physical space relying on this being the only condition accessible to experience (p.124). Whether one includes dreams into the definition of experience is an open question.

In four-dimensional space, a sphere can be turned inside out without cutting, and knots can be undone also without breaking the continuity of the cord or moving its ends. Also, as by definition, the fourth dimension 'is directed' away from all existing ones, a point starting at the centre of a three-dimensional sphere could move along this new axis without approaching any portion of the surface of the sphere, whilst moving away at the same rate from all points on this surface (Manning, 1909, p.23).[4]

Mathematically, a four-dimensional space is one where a point requires four coordinates in order to be uniquely specified. There are several ways of conceptualising the idea of a fourth dimension, with a common interpretation being that of physical space as we perceive it, plus the dimension of time as the fourth. Among these, the most common is the so-called Minkowski spacetime (non-Euclidean), which formed the basis for Einstein's theories of relativity. This is a particular kind of four-dimensional space, one where the dimension of time has a special status. Notably, in this formalisation, measurements along the fourth dimension are made using imaginary numbers, which highlights the lack of homogeneity among the dimensions. Time includes all things at once, while in space no two things can share the same point. Interactions occur through movement, whereby space is temporalised and time is spatialised. Also, movement along the dimension of time can occur (or at least be experienced) in only one direction. As Paty (1998) emphasises, space and time are abstract concepts which are constituted on the bases of the experience of our senses, and of scientific reasoning (p.2). He offers a good historical overview of possible explanations for the prevalence of three-dimensional space (from the importance of number 3 in ancient times to physiological considerations related to perception), as well as for the purely symbolic (in Lacanian terms) necessity of higher dimensionality, as it emerges from the development of mathematical modelling of the physical universe. In this sense, higher dimensions are not 'natural' observations, but, rather, are hypothesised structures given legitimacy through the theoretical developments which rest upon them. In other words, once the existence of the fourth dimension is presupposed, mathematical problems hitherto unsolvable yield solutions that are not just theoretical, but that also have practical relevance. This is a transformation in terms of the power of purely abstract mathematics of the same nature with the one produced by the introduction of imaginary and complex numbers. The square root of minus one cannot be observed anywhere, yet postulating its existence has made it possible to find solutions to many practical problems that could not be worked out otherwise. Likewise, with the hypothesis of the Freudian unconscious, to which over 100 years of clinical work attest.

Paty (1998) notes the fact that this new understanding of spacetime as a single four-dimensional structure, this relatively new mathematical ontology, has not acquired the same status of popularity as that of three-dimensional space, for which, as Poincaré (1912) stresses, we have a 'net intuition' (p.504). Each place in space can only be perceived inexactly (Crilly, 1999, p.13). In other words, experience does not prove to us that space actually has three dimensions; it merely shows that we find it comfortable to attribute it this number (Paty, 1998, p.13).

In the familiar system of spatial representation using Cartesian axes, the common point between these is the origin, marked conventionally by zero. That is at the level of the Symbolic. But at the level of experience, the prime reference point for spatial orientation in relation to external physical objects is the body (Garella, 2012, p.79). This is the point of view expressed by Poincaré (1912) himself, who is clear that space only exists relative to a certain initial position of the body.

When I refer to the fourth dimension here, I am not referring to time, but I mean it in purely spatial terms,[5] that is to say, as a four-dimensional space departing from the standard view of three-dimensional Euclidean space, a development in geometry established as a valid option in the early 1900s, just as psychoanalysis itself was becoming established through Freud's work. This came to the fore through what is known as the Kaluza-Klein theory (Greene, 2000) introduced in the 1920s, offering the possibility of the unification of disparate fields of physics, something that was not just Einstein's dream. The fact that it predicted the existence of a particle not yet observed made this approach less than popular for a while, until the 1970s, when it was picked up again and built upon in the context of string theory developments (Binétruy, 2003, p.59). The way physics understands space is still evolving. As things stand, the prominence of the three dimensions of space as we perceive it does not rule out the possible existence of further spatial dimensions, and it is not a small detail that superstring theory, which posits the existence of numerous other dimensions, stands out among theories of physics in that it is not plagued by internal contradictions or singularities[6] (Barrow, 1999, pp.184–185).

The question of the actual existence of the fourth dimension has been of great interest over time, not only to mathematicians, but also to philosophers and physicists, in their pursuit of understanding the nature of the space we inhabit, as well as to spiritualists, who welcomed this notion as the answer to the whereabouts of the spirit world. It is one main reason for which Lacan dismissed the idea as associated with 'science fiction' (Lacan, 2020, p.322).

In a more considered take, Gumaer (1905) distinguishes between mathematical and physical reality:

> If existence means that the intellectual idea of a thing can be formed, and that this idea shall not lead to contradictions with other well established ideas and with the results of our experience, then it may be said that four-dimensional space does exist. If, on the other hand, existence is taken to mean objective or actual reality, all we can say about it is that we do not know.
>
> (p.170)

5.2 The unconscious as four-dimensional space

What is the relevance of all this to the unconscious? As the title of this section indicates, what I submit for consideration is not that the idea that the unconscious is located *in* some impenetrable hyperspace, but that its structure is that of a four-dimensional space. More specifically, I am focused on the relevance of the interplay between three- and four-dimensional space, namely, on the absence of continuity and coherence at the level of experience, which can be understood as the experience of four-dimensional processes which are collapsed or reduced to the habitual perception of three-dimensionality, and thus unavoidably fragmented and seemingly disjointed.

In this sense, the inaccessibility of the unconscious is structural. No matter how hard we try, we cannot apprehend it as a whole, as it is not a secret chamber that could be located, sooner or later, and entered into. It is only through intermittent encounters that we can perceive its productions as fragments, but which are parts of a single whole whose coherence resides in a space of higher dimensionality. These intermittent encounters can be understood as instances of intersection between the three-dimensional space we inhabit and the four-dimensional unity of the unconscious as a kind of personal hyper-object. The work of analysis, through speech and its interruptions, is the closest thing we have got to a way of engaging with this dimensional disparity. Lacan (1979) offers a recognition of this process when he refers to the intermittent pulsation-like opening of the unconscious. More specifically, he conceptualises the unconscious itself as a cut that opens and closes (Lacan, 1979, p.43), a description that is fully consistent with a sequence of recurring encounters with an entity that occupies in its entirety a space of higher dimensionality. The easiest way to relate to this is through dreams, with the dreamer finding themselves both at home and somewhat uneasy in their own psychic productions. Lacan's reading of Freud's mapping of the unconscious in the *Interpretation of Dreams* is as a spatial locus, 'an immense display, a special spectre, situated between perception and consciousness' (Lacan, 1979, p.45), the other location, or space or scene in Freud's words, '*einer anderer Localitat*' (Lacan, 1979, p.56).

If we accept this hypothesis of the unconscious structured as a four-dimensional space, then condensation is no longer to be regarded as the result of the dream work of encryption, but merely as a structural consequence of projection from a higher to a lower dimension: overlaps will occur and details will be lost. Elements that are juxtaposed in four-dimensional space appear as overlapping in three-dimensional space. An easily understood equivalent process to consider is that of our own two-dimensional shadows, which lack colour and depth and in which a mere contour around some undifferentiated compactness is all that remains. Nevertheless, while the shadow is in direct relation to the body, it is not the body. Shadows cast from different angles may appear different, yet they refer to the same entity whose unity and totality cannot be captured in any single representation.

Equally, repetition is no longer simply the revisiting of a pattern or situation, but the totality of one single instance of the same unconscious, four-dimensional phantasy, which is experienced in recurrent form in three dimensions, with time taking the place of the fourth dimension at the level of experience. Each instance is a kind of three-dimensional slice through a structure akin to a four-dimensional space. The work of analysis is to re-establish the continuity between these instances and then to query the very four-dimensional structure at play. This is why the cut of Lacanian practice is so important, clinically, as its mathematical equivalent is the one operation capable of transforming the very structure of the space upon which it operates.

Bursztein (2008) argues that space in psychoanalysis has nothing to do with physical space and that the question of time can explained as variations of space. Indeed, he stresses that '[N]on Lacanian psychoanalysts emphasise the dimension of time, while the Lacanians stress the dimension of space' (2008, p.76). Except this demarcation by orientation is not absolute.

From the non-Lacanian post-Freudian tradition, Bion (1994) posits the existence of 'a mental multi-dimensional space of unthought and unthinkable extent and characteristics' (p.313), which includes a domain of thoughts without a thinker alongside thoughts with a thinker.

As we have seen, when (mathematically) projecting from a higher-dimensional space into a lower-dimensional one, **overlaps** occur: what is separate or distinct in reality can appear as identical, so false identifications are at stake. This is consistent with the use of the term projection in post-Freudian psychoanalysis, in particular in the Kleinian tradition.

Equally, a higher-dimensional object can appear like a sequence of disparate lower-dimensional objects in the lower-dimensional space, generating the appearance of **repetition**. A common occurrence of this kind, physically, is the sequence of two-dimensional plates obtained, for instance, from a brain scan. They all look similar, when placed in two dimensions, as if something were being repeated. Instead, each slide is a picture of the magnetic resonance cut through the same three-dimensional brain, at various planes of intersection. Psychoanalytically, repetition is central to each clinical presentation and, as transference, to each clinical encounter. Lacan (1979) singles out repetition and transference as the fundamental concepts of psychoanalysis, alongside the unconscious and the drive. According to Chemama and Vandermersch (2009), it is repetition that knots together the other three concepts, as it operates as the stumbling block of the unconscious, the pivot for transference and the very principle of the drive (pp.498–499). Repetition is also the notion that took Freud to the recognition of the death drive which, in Lacan's interpretation, became understood as that which is impossible to symbolise, namely, the encounter with the Real (Chemama and Vandermersch, 2009, p.501).

Lacan is careful to distinguish between repetition that pertains to the encounter with the Real, specifically as the missed encounter (*tuché*), and repetition that is in the nature of the insistence of the signs, of the Symbolic order (*automaton*) (Lacan, 1979, pp.53–55). This is, in essence, the return of trauma, which, in turn, is an encounter with discontinuity expressed by both Freud and Lacan in spatial terms. What is at stake, says Lacan, is an 'experience of rupture [...] which forces us to posit what Freud calls *die Idee einer anderer Lokalität*, the idea of another locality, another space, another scene, the between perception and consciousness' (p.56).

If we accept the hypothesis that the unconscious is structurally a four-dimensional space, then it is easy to understand why it is not immediately accessible as such, but rather it only lends itself to a fragmented, stuttering kind of access, via encounters which have the appearance of repetition. The impossibility of continuity is not in the discontinuous nature of the unconscious itself, but rather a consequence of the limited nature of access from lower- to higher-dimensional spaces, structurally. Lack, as well as trauma, therefore, emerges as a consequence of the limitations to moving from a four-dimensional space of the mind to the three-dimensional one inhabited by the body and experience. No one-to-one mapping is ever possible between the two. This additional dimensionality is one that emerges with the signifier, yet it cannot be accessed directly by language and speech. Witness the struggle and anguish of the Square, when he tries to capture into one-dimensional words all that he had experienced outside the confines of his familiar physical space.

The prohibition that confronts the Square in the story of *Flatland*, encountered in Chapter 4, is merely a veil. Had the authorities allowed him to speak or write freely about the spaces of higher dimensionality he had accessed, the intrinsic nature of his space would have prohibited him from bringing those spaces into Flatland. Higher does not fit into lower, but unlike the case of numbers, one does not end up with a fraction of the total, but a set of disparate fragments. Like in the case of the Oedipus complex, prohibition is the veil to impossibility of satisfaction (Verhaeghe, 2009).

What seems to be a sequence of recurring experiences, made chronological by speech, could be more coherently understood as parts of the same one unconscious construction – the fundamental phantasy. Phantasy, in turn, is in a direct relation to the real: 'The real supports the phantasy, the phantasy protects the real' (Lacan, 1979, p.41); it is the veil. The construction of the fundamental phantasy is one which can be reduced to a minimal axiom, which is impersonal. The Lacanian view of analysis is one where the subject traverses it, to move into a position where new phantasies can arise. The term 'traversing' has strong spatial connotations, which raises the question of what may be the topology of the pass. More on this in Part III.

Lacan dismisses the idea of three-dimensional space, as he insists on conceiving of space as a set of relationships rather than as a place. In his words, 'There is no real space. This is a purely verbal construction which has been spelt out in three dimensions' (Le sinthome, 96, 10 February 1976, cited from unprinted source by Cléro (2002, p.41)). He refers to this as an additional dimension, the one of what is spoken, of what is said (dit-mension) in *L'Etourdit*. For him this is a dimension of the impossible (Lacan, 1972, p.476). Lacan places the subject of the unconscious 'in a sort of internal beyond' (1997, p.123). He builds thus upon Freud's (1915) argument that 'internal objects are less unknowable that the external world', with the correction of internal perception (p.171) that the recognition of the unconscious makes possible.

In this sense, we could say that the so-called 'internal' world is four-dimensional, while the perceived external world is three-dimensional and the body makes a sort of cut between the two.

5.3 Dimensions in the clinic

It is a common occurrence to hear patients say: 'part of me feels that way, part of me thinks that...', in a manner that indicates that various parts of oneself coexist while they are irreducible, incompatible and often impossible to reconcile. Such difficulty with ambivalence speaks to an emotional encounter with the impossibilities of fitting higher dimensions into a space of lower dimensionality. In other words, this is consistent with the hypothesis that the unconscious has a four-dimensional structure, whilst experience occurs in a three-dimensional space. More than one thing can be true at the same time, because in the unconscious two things can occupy the same position at once. Indeed, analysands often comment on how the realisation that 'both can be true' transforms their experience of their own conflict in such a way as to facilitate movement towards more freedom and ease. The experienced split is translated as continuity in another dimension, where new movement becomes possible.

This subtle interplay between dimensions is precisely what Lacan rested his optical schema of Seminar I upon, implicitly, where the coherence of the junction between the Real and the Imaginary is shown to depend upon the subject taking a certain symbolic position[7] (Lacan, 1991a).

In the experiment Lacan uses to illustrate this point, the image captured in a curved mirror, of the reflection of an actual vase and the inverted image of actual flowers hanging under the vase, can only be perceived as the familiar picture of 'flowers in a vase' when the observer is at a certain position in relation to all the component elements of the optical construction.

The main point of this schema is to illustrate the distinction between registers of experience, and although at this stage in his work Lacan focuses explicitly on separating the Symbolic from the Imaginary, the notion of the Real is implicit as something that escapes the other two registers. The image gains coherence at the point of interplay between these dimension-like registers. Specifically, he stresses that in clinical encounters, the patient addresses the analyst on the level of the Imaginary as another person just like themselves, and makes demands from this other person, while on another level, that of the unconscious, the one addressed or engaged with in the analytic process is a different kind of Other, a symbolic construction that holds together the mystery of the patient's own being – an Other to be appeased, seduced, obeyed, defied, escaped. It is thorough the very work of analysis that the relationship to this unconsciously constituted Other and their demands can be modified and therefore new possibilities of organising one's experience of being can be accessed, away from the suffering that prompted the search for analysis in the first place.

Whilst the Imaginary axis is one predicated on the presumption that we know who we are and who the other is, the Symbolic dimension is an axis of otherness. This is where some articulation around the question of 'Who am I?' can begin to be sketched. The Other is more of a location than a person; it is the other place where we dream and from where we seek answers, the place where our desire originates. This is already hinting at another dimension where the unconscious is situated, on the side of the Other, of oneself yet outside of it, the internal beyond (Lacan, 1997).

Indeed, Lacan's extensive criticism of the prevailing IPA mode of working, and specifically of Melanie Klein's work, returns time and again to the failure to distinguish both theoretically and in the clinic between the registers of the Symbolic and the Imaginary. Her confounding the two can be understood as a failure of perspective, but also as a collapsing of dimensions.[8] It is also by de-confounding the imaginary and symbolic aspects of transference that Lacan is able to reconcile the apparently conflicting views between transference as vehicle of treatment and as a barrier to the progress of analysis.

In terms of Lacan's L-schema introduced in 1955 (Lacan, 1991b) and refined in subsequent years, the patient enters analysis with an experience predominated by and confined more often than not to an awareness of the Imaginary dimension. Lacan represents this by a line, which geometrically constitutes a space that only allows for movement along it, back and forth, towards or away from the other. By emphasising the Symbolic, the analyst introduces a new axis and thus opens up the space of this line into a plane. In adding a new dimension, the analyst creates a space where movement in another direction becomes possible, taking away from the pressure on the Imaginary to answer questions beyond its remit, and giving access to the more relevant and potent realm of language and speech, and, through that, to the Unconscious. As Lacan clarifies, the one who arrives to speak in analysis is stretched over the four corners of the schema: as the subject S, 'his ineffable and stupid existence', as o' – his objects, as o – his ego, and as the other, A – 'the place from which the question of his existence may be posed for him' (Lacan, 1958, p.459). The aim of psychoanalysis becomes true speech, which links S to A, undermining thus the Imaginary axis (Leader, 2003, p.184).

In mathematical terms, this difficulty with accurate perception in projections is also central to knot theory (more of which in Chapter 8), and the classification of knots. For now, we can illustrate a key point by imagining a rope knot floating in the middle of an art studio (Arroyo, 2016); it is easy to see how the drawing (i.e. two-dimensional projection) of each artist in the circle of easels around it would be different. How is one to determine, by only looking at the drawings, that they were all depicting one and the same knot?

This question of classification is a common experience in the Lacanian clinic, where the psychic structure for each patient is something that can only be recognised with any degree of confidence after exploration from various angles. Sometimes this is easier to establish, when the knot of the structure is encountered from a perspective that supports clarity. The knot of the psychic structure is the same; the position of the analyst is what becomes refined by attending to the speech of the patient.

If all we can encounter is not the unconscious as such, but its lower-dimensional shadows, then, like poetry, psychoanalysis can only hope to decipher something about the layers of these shadows, whilst recognising their enduring other-dimensional impenetrable mystery.

If we accept the hypothesis of the unconscious structured like a four-dimensional space, the main clinical implication is that all its formations can be understood as elements of a set in a space of inaccessible dimensionality. This approach also explains why the pursuit of reducing experience to a coherent narrative, in language, is both inaccurate and destined to fail – a reality of which Lacanian analysts are much more aware than those of other orientations. Instead, it is not just possible, but also consistent with the nature of what is addressed, to allow for gaps and to work around them. The recognition of points of discontinuity operates both at the level of experience in relation to the unconscious and at the level of language in relation to experience. Collapsing everything into sense would deny the very nature of the processes at play, not in terms of the fragmentation of the unconscious itself, but with regard to the impossibility of translating into any kind of completeness and continuity any transmission from a higher to a lower dimension.

If we understood the difficulty with reaching the unconscious as such, in a way that could produce direct change, as a question of inaccessibility, our approach would be to find a better route to the 'far away place', a kind of shortcut. If, however, we recognise this same difficulty as one produced by structural impossibility (in carrying over any entity from a space of higher dimensionality to a space of lower dimensionality), then the work of analysis takes an entirely different orientation, as it accompanies the subject on a journey of creating a way of being with impossibility itself, rather than blaming themselves for it or trying to resolve it, in a way that is destined, structurally, to fail.

The question of impossibility itself can be pursued further, by challenging the very notion of dimensionality, or by exploring the properties of spatial structures that incorporate impossibility itself.

5.4 No dimensions and the inscription of impossibility

In his relentless pursuit of an understanding of the nature of infinity, in the 1870s, Cantor posed a question that challenged the very notion of dimension, and his work on this led to his recognition as the 'true father of dimension theory' among mathematicians (Crilly, 1999, p.1). His interest was in establishing whether there was a two-way one-to-one relationship between all the points on a side of a square, and all the points inside the square. This is known in mathematics as a bijection, a function whereby each element belonging to one set is paired with exactly one element from another set. If a one-dimensional side could be seen, in this sense, to be equivalent to a two-dimensional surface, the very notion of dimension would become redundant.

His proof that this was the case relied on introducing discontinuity everywhere (Charraud, 1994, p.229). That is to say, whilst the spaces in question remained continuous, the mapping between them was not (Dauben, 1975). His remarkable line of thought abandoned entirely the Imaginary, placing the question fully in the domain of the Symbolic. The relationship Cantor considered was no longer one between geometrical shapes, but between numbers in the interval [0,1], a space which contains an infinity of numbers, as we have seen in Chapter 3 (Rosset, 2013, p.87). The move away from geometry as we know it to mere relationships between numbers constituted a stepping away from visual representations to pure symbolism. Cantor himself was taken aback by his own result, and he is known to have exclaimed: 'I see it, but I do not believe it' (Crilly, 1999, p.4). He was so troubled by this that he set out to prove his own results as wrong, a derivation which was later found to be false (Crilly, 1999, p.9).

Cantor established this two-way one-to-one relationship as follows: for any number on the segment [0,1], he separated numbers from alternate decimal places into two distinct numbers, which became the planar coordinates in the square of side 1. For instance, 0.752438 was separated into two distinct coordinates, by grouping alternate decimal places from 0.**7**5**2**4**3**8 into 0.723 (digits in bold) and 0.548 (remaining, alternate digits) (see Figure 5.1).

This transformation amounts to an operation of encryption, as numbers are unravelled and put together according to a particular key established by him.[9] In this scenario, numbers do not operate as measures of magnitude, but simply become mere coordinates, a way of uniquely identifying points; they become simple names, but ones upon which certain operations are possible.

The relationship he established between the two points located in the unidimensional and two-dimensional spaces was such that only one point in one would correspond to only one point in the other. This established a counter-intuitive equivalence between a unidimensional and a two-dimensional object. As Vivier (2004) points out, this amounts to saying that a one-dimensional object and a two-dimensional object have the same number of elements (p.122). It is important to note that, although bijective, this function is not continuous, which means that the correspondence identified by Cantor does not amount to saying that a segment

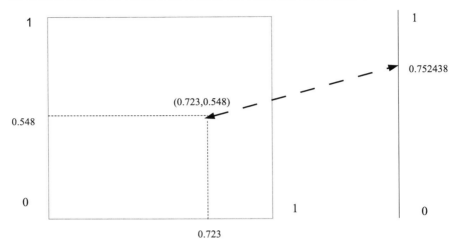

Figure 5.1 Bijective mapping between segment and square.

and a square are topologically equivalent. In other words, they are not homeomorphic:[10] it is not possible to continuously deform one of them and turn it into the other (the invariance of dimensions theorem, Alexandroff, 1961, p.1). Instead, this is a question of mapping one infinity onto another.

Cantor's proof brought to the fore the inherent lack of continuity not in the spaces themselves, but in the movement between them. This is consistent with the discontinuities in translation between the space of the unconscious and that of experience, as well as between experience and language.

Later on, Cantor generalised this finding to figures of any number of dimensions, all of which could be put into a one-to-one correspondence with one-dimensional continuous lines (Crilly, 1999, p.5). His work and that of mathematicians of his era gave a devastating blow to the coordinate concept of dimension, and called for a new way of thinking about space, opening the door to new elaborations about set theory and topology. The consequences of his findings are put in striking terms by Kasner and Newman (1940): 'a line segment one-millionth of an inch long has as many points as there are in all three-dimensional space in the entire universe' (p.56). The way in which Cantor became able to think about space was entirely abstract, counter-intuitive and fully rigorous.

One of the many consequences of Cantor's provoking work was the development of topology itself, as a version of rubber-sheet geometry where space became formulated in radically new ways. It is this particular branch of mathematics that held much of Lacan's interest in the latter part of his work, in terms of formulating the essence of the unconscious, and it is to this that we turn in Part III, where we extend this exploration from the question of inaccessibility to that of impossibility.

Notes

1 E.g. the dream of Irma's injection (Freud, 1900, pp.106–121).
2 I am aware that this intuitive approach to dimensions deviates from the pure mathematical definition of dimensions independent of any notion of movement (see Crilly, 1999).
3 In algebraic terms, there is no limit to the number of dimensions a mathematical space can have, but any expansion beyond four dimensions remains outside the scope of this book.
4 We will come to examine another spatial configuration with attributes that resonate with this one when we consider the cross-cap in Chapter 7.
5 This is consistent with the definition of time arrived at by Bursztein (2017), namely, as a cut in space (p.72). More specifically, he argues that such cuts are realised subjectively as Möbian structures, such that they account for both the sequential nature of lived experience and the timelessness of the unconscious.
6 Singularities are points where the laws of physics break down and the usual concept of space and time lose their meaning.
7 Almost two decades later, for both the French and later the English edition, the cover of Lacan's widely read seminar on *The four fundamental concepts of psycho-analysis* reproduces a representation of anamorphosis, which consists of a distorted projection of perspective that requires the viewer to look from a specific position, use some special devices or both, in order to view the image as something recognisable. Specifically, this is a reproduction of Holbein's *The Ambassadors*, of 1533, where the skull reminder of death, juxtaposed to the many riches that surround the men, can only be seen for what it is only from a particular vantage point.
8 A good illustration of this notion is the case of familiar representations of celestial maps of named constellations. In their two-dimensionality, these are misleading about the extent to which stars that appear next to each other on the map are actually separated by unfathomable distances along the dimension that is flattened by the map, namely that of depth. See, for instance, a virtual trip around Orion that can be viewed at https://www.youtube.com/watch?v=lD-5ZOipE48 https://www.youtube.com/watch?v=lD-5ZOipE48
9 Thanks to Serban Sovaiala for help with clarifying this aspect.
10 The first to prove dimensional invariance, in the early 1900s, was L.E.J. Brouwer, a prominent topologist with an interest in intuitionism, whereby mathematics is regarded as purely the result of the constructive human mental activity of rather than the discovery of fundamental principles claimed to exist in an objective reality. He showed that two spaces of dimensions m and respectively n are homeomorphic if and only if $m = n$.

References

Alexandroff, P. (1961). *Elementary concepts of topology*. New York: Dover Publications.
Arroyo, A. (2016). *Les nœuds sauvages, Calendrier mathématique 2016*, Strasbourg: Presse Universitaire de Strassbourg.
Barrow, J.D. (1999). *Impossibility: The limits of science and the science of limits*. London: Vintage.
Binétruy, P. (2003). Les nouvelles dimensions de l'Univers. In Cohen-Tannoudji, G. and Noël, È. (eds.). *Le réel et ses dimensions*. Editeur: EDP Sciences, pp.57–67.
Bion, W.R. (1994). *Cogitations*, Bion, F. (ed.). London: Karnac.
Bursztein, J-G. (2008). *On the difference between psychoanalysis and psychotherapy*. Paris: Nouvelles Etudes Freudiennes.

Bursztein, J-G. (2017). *L'Inconscient, son espace-temps: Aristote, Lacan, Poincaré.* Paris: Hermann.
Camp, B.H. (1905). The fourth dimension algebraically considered. In Manning, H.P. and Mitchell, S.A. (1909) (eds.). *The fourth dimension simply Explained: A collection of essays selected from those submitted in the Scientific American's prize competition*, University of Michigan University Library, pp.118–124.
Caussanel, M-L. (2011). *Psychanalyse et topologie: Introduction aux dimensions négatives.* Exposé au Kafemath.
Chambers, G.G. (1909). The meaning of the term "fourth dimension". In Manning, H.P. and Mitchell, S.A. (eds.). *The fourth dimension simply Explained: A collection of essays selected from those submitted in the Scientific American's prize competition*, University of Michigan University Library, pp.201–210.
Charraud, N. (1994). *Infini et inconscient: Essai sur Georg Cantor.* Paris: Anthropps.
Charraud, N. (1997). *Lacan et les mathématiques.* Paris: Anthropps.
Chemama, R. and Vandermersch, B. (2009). *Dictionnaire de la psychanalyse.* Paris: Larousse.
Cléro, Jean-Pierre (2002). *Le vocabulaire de Lacan.* Paris: Ellipses.
Crilly, T. (1999). The emergence of topological dimension theory. In James, I.M. (ed.) *History of topology.* Oxford: Elsevier, pp.1–24.
Cutler, E.H. (1909). The fourth dimension absurdities. In Manning, H.P. and Mitchell, S.A. (eds.). *The fourth dimension simply Explained: A collection of essays selected from those submitted in the Scientific American's prize competition*, University of Michigan University Library, pp.60–69.
Dauben. J.W. (1975). The invariance of dimension: Problems in the early development of set theory and topology. *Historia Mathematica*, 2, pp.273–288.
Downing, D. (2009). *Dictionary of mathematics terms*, 3rd ed. New York: Barron's Educational Series.
Evans, D. (1996). *An introductory dictionary of Lacanian psychoanalysis.* London: Routledge.
Fitch, G.D. (1909a). An elucidation of the fourth dimension. In Manning, H.P. and Mitchell, S.A. (eds.). *The fourth dimension simply Explained: A collection of essays selected from those submitted in the Scientific American's prize competition*, University of Michigan University Library, pp.43–51.
Fitch, G.D. (1909b). Non-Euclidean geometry of the fourth dimension. In Manning, H.P. and Mitchell, S.A. (eds.). *The fourth dimension simply Explained: A collection of essays selected from those submitted in the Scientific American's prize competition*, University of Michigan University Library, pp.52–59.
Freud, S.(1900). The interpretation of dreams. SE4 & SE5.
Freud, S. (1915). The unconscious. SE14, pp.161–215.
Freud, S. (1930). Civilisation and its discontents. SE21, pp.57–145.
Freud, S. (1950 [1895]). Project for a scientific psychology. SE1, pp.281–397.
Gabriel, M. (2016). New realism. *Philosophy Now*, issue 113, April/May 2016.
Garella, A. (2012). Exploration of the unconscious: Some considerations on space, the object and the process of knowledge in psychoanalysis. *The Italian Psychoanalytic Annual*, 6, pp.73–89.
Greene, B. (2000). T*he elegant universe: Superstrings, hidden dimensions and the quest for the ultimate theory.* London: Vintage Books.

Gumaer, Percy Wilcox (1905). The true and the false in the theory of four dimensions. In Manning, Henry P. and Mitchell, S.A. (1909) (eds.). *The fourth dimension simply Explained: A collection of essays selected from those submitted in the Scientific American's prize competition*, University of Michigan University Library, pp.163–171.
Kasner, E. and Newman, J. (1940). *Mathematics and the imagination*. New York: Dover Publications.
Lacan, J. (1949). The mirror stage as formative of the I function as revealed in psychoanalytic experience. In Fink, B. (ed.) (2002). *Écrits*. London; WW Norton, pp.75–81.
Lacan, J. (1951). Some reflections on the ego. *International Journal of Psychoanalysis*, 34, pp.11–17.
Lacan, J. (1958). On a question prior to any possible treatment of psychosis. In Fink, B. (ed.) (2002). *Écrits*. London: WW Norton, pp.445–488.
Lacan, J. (1972). L'Étourdit. In Lacan, J. (2001). *Autres ecrits*. Paris: Éditions du Seuil.
Lacan, J. (1979[1973]). *The four fundamental concepts of psychoanalysis*. London: Penguin Books.
Lacan, J. (1991a[1953]). *The Seminar of Jacques Lacan: Book 1, Freud's Papers on Technique*. New York: W. W. Norton & Company.
Lacan, J. (1991b[1978]). *The seminar. Book II. The ego in Freud's theory and in the technique of psychoanalysis, 1954–1955*. London: WW Norton.
Lacan, J. (1997 [1981]) *The seminar. Book III. The psychoses, 1955–1956*. London: WW Norton.
Lacan, J. (2020). *L'identification: Séminaire 1961–1962*. Éditions de l'Association Lacanienne Internationale. Paris: Publication hors commerce.
Leader, D. (2003). The Schema L. In Burgoyne, B. (ed.), *Drawing the soul: Schemas and models in psychoanalysis*. London: Routledge, pp.172–189.
Manning, H.P. (1909). Introduction, in Manning, H.P. and Mitchell, S.A. (eds.). *The fourth dimension simply Explained: A collection of essays selected from those submitted in the Scientific American's prize competition*, University of Michigan University Library, pp.7–41.
Matte Blanco, I. (1975). *The unconscious as infinite sets* (1975). Aylesbury: Duckworth
Matte Blanco, I. (1988). *Thinking, feeling, and being: Clinical reflections on the fundamental antinomy of human beings and world*. London: Routledge.
Nasio, J-D. (2010). *Introduction à la topologie de Lacan*. Paris: Petit Bibliothèque Payot.
Paty, M. (1998). Les trois dimensions de l'espace et les quatre dimensionsde l'espace-temps. In Flament, Dominique. *Dimension*, dimensions I. Paris: Fondation Maison des Sciences de l'Homme, pp.87–112., Série Documents de travail.
Poincaré, H. (1912). Pourqui l'espace a trois dimensions. *Revue de métaphysique et de morale*, 20(4), pp.483–504.
Riemann, D. (1854). On the hypotheses which lie at the foundations of geometry. In Smith, D.E. (ed.). (1959). *A source book in mathematics*. New York: Dover Publications, pp.404–425.
Rosset, J.P. (2013). Le réel entre mathématiques et psychoanalyse. Seminaire de psychanalyse 2012–2013, ALI Alpes-Maritimes-AEFL, pp.83–95.
Rovelli, C. (2014). *Seven brief lessons on physics*. London: Allen Lane, Penguin Books.
Sogyal, R. (2002). *The Tibetan book of living and dying*. London: Rider.
Verhaeghe, P. (2009) *New studies of old villains: A radical reconsideration of the Oedipus complex*. New York: Other Press.
Vivier, L. (2004). *La topologie: L'infini matrisé*. Paris: Le Pommier.

Part III

The unconscious as domain of impossibility

The logical movement from infinity to impossibility is not exclusive to psychoanalysis. This was the fate of thought in several disciplines in the 19th century: the idea of infinite knowledge that could be, conceivably, accessed by a mind great enough (Laplace's superbeing - see Barrow, 1999, pp.48–50), was discarded in favour of a formal recognition of impossibility, which emerged as a finding of science in itself. The culmination of this was in Godel's incompleteness theorem, which showed that arithmetic cannot prove its own completeness and consistency - more formally, that 'a rigid logical system will contain true propositions that cannot be proved to be true' (Downing, 2009, p.146). This striking result opened the door to new insights, as a number of paradoxes became formulated, and impossible problems in mathematics were identified. The pursuit of the unknown lead thus to a recognition of the unknowable and of the impossible.

As Rosolato (1978) observes, the prevailing course of action with the unknown is to keep it at bay, with research regarded as a quasi-spatial mission of conquest that expands the boundaries of what is known (p.273). The notion of impossibility as an absolute limit to what science can decipher was not and is not a popular one, and some reject it as a depressing barrier to knowledge. The vibe in this century is still on the optimistic side, as Du Sautoy's (2016) book illustrates. Formal, rigorous thought about impossibility itself is difficult to find outside of the realm of psychoanalysis. Again, it is Lacan who addresses this most directly in his later work, where he examines the topology of the nature of the unconscious and of human experience. To him, 'the unconscious is the impossible' (Greenshields, 2017, p.23). Lacan's recourse to topology is not accidental, as this is the only branch of mathematics that takes on the task of formalising impossibility. Through its focus on discontinuity, topology offers a unique understanding of spaces where certain "areas" are inaccessible not because they are tricky to get to or because it takes a long time to arrive at them, but because they simply do not exist.

Chapter 6 considers the key notions involved in formulating impossibility in mathematical terms, then moves on to an exploration of the relevance of these developments to the central idea of psychoanalysis, namely that of the Oedipus complex. Chapter 7 introduces key topological ideas and explores their relation to psychoanalysis as formalised by Lacan and some of his followers, while Chapter 8 explores notions of knot theory, a subfield of topology of particular relevance to clinical work.

We begin with an examination of and engagement with established rigorous mathematical formulations of impossibility in spatial terms. These allow us to take further the proposed reading of the unconscious as space, and offer structural foundations for equally rigorous psychoanalytic formulations around impossibility, such that what is at stake is neither ignored, nor papered over, but recognised and circumscribed. Part IV follows this through into the realm of clinical implications and analytic technique.

References

Barrow, J. D. (1999). *Impossibility: The limits of science and the science of limits*. London: Vintage.
Du Sautoy, M. (2016). *What we cannot know*. London: 4th Estate.
Downing, D. (2009). *Dictionary of mathematics terms*, 3rd ed., Barron's Educational Series.
Greenshields, W. (2017). *Writing the structures of the subject: Lacan and topology*. London: Palgrave Macmillan.
Rosolato, G. (1978). *La relation d'inconnu*. Paris: Galimard.

Chapter 6

Structures of the impossible

In his reception lecture which marked his entry to the Romanian Academy, at the end of a long career as a mathematician, Solomon Marcus focused on the discord between what becomes intelligible through mathematics and what can be perceived directly, and stressed that – like poetry – mathematics transgresses the locus of daily existence into the counter-intuitive and paradoxical aspects of existence (Marcus, 2014, p.10). Much of his work addressed the exploration of such paradoxes as 'pathological' aspects in sets and functions, and he devoted years to teaching mathematical linguistics. He particularly recognised one of the functions of mathematics as that of providing the means for understanding the mind (Marcus, 2014, p.35).

Mathematically, the impossible is expressed in terms of discontinuity, which carries with it the radical exclusion of any chance of movement from certain positions to others. In most general terms, these are points of singularity, that is to say, points at which a given mathematical object is not defined or ceases to be 'well-behaved',[1] instances where certain positions cannot be arrived at structurally. They are not just lost or difficult to access, but fundamentally unavailable.

Psychoanalytically, impossibility is not a formal concept as such, but rather a notion implied in the castration complex which derives, in turn, from the Oedipus complex, as introduced by Freud (Laplanche and Pontalis, 1988, pp.56–59). For Freud, this complex constitutes the bedrock against which analysis stumbles and crashes, the biologically-imposed limit to what analysis can reach and therefore transform (Freud, 1937). Among post-Freudians, it is Lacan who clarifies the link to impossibility, by placing castration away from biological concerns and into the realm of the symbolic, as an operation which determines the structure of the subject. In Lacanian terms, castration implies renunciation of both being and having the phallus, an operation which institutes phantasy as the engine and sustainer of desire (Chemama and Vandermersch, 2009, p.97). This goes beyond the incest taboo and submission to the prohibition of the father to seek satisfaction in the mother, into the territory of assuming the lack as engine of desire. Some post-Lacanians view castration as synonymous with existential lack itself, an expression of the human condition.[2] As we are going to see, the satisfaction that is forbidden is not merely rendered inaccessible by the law, but intrinsically impossible in the

DOI: 10.4324/9781003479284-9

sense that it does not actually exist – which is the very trajectory of our consideration and formulation in spatial terms.

An introduction to ways in which mathematics formalises something about impossibility is followed by an overview of the landscape of possible outcomes in the subjective encounter with castration in this sense.

6.1 Mathematical representations of impossibility

In mathematical terms, impossibility relates to instances or locations where movement or transformation available elsewhere cannot be accessed at all. These are, in essence, instances of discontinuity, and the primary connotation is always a spatial one, with singularities defined as points of impossibility in an explicitly defined or implicit space. For so many years, time and space were thought of as continuums of points, yet this could be recognised as no more than a convenient assumption that is desirable for simple mathematics (Barrow, 1999, p.225). The development of quantum physics rests largely on doing away with such situations, that is to say, getting rid of zero to get rid of singularities. As Seife (2000) puts it, '[T]he Theory of Everything is, in truth, a theory of nothing' (p.192). We can recognise in this the Lacanian understanding of the subject as structured by its relation to lack. The theory of the subject and their being is the theory of lack and impossibility.

Whilst topology offers the most elaborate understanding of the relevance of what is at stake, as Chapter 7 shows, we begin this exploration with a familiar instance of impossibility expressed in mathematical terms: division by zero. In terms of calculus, this can be best illustrated through the real function $f(x) = 1/x$, where x is any number from the set of real numbers (hence the function is real in the mathematical rather than Lacanian sense). For $x = 0$, the function seems to 'explode' to plus or minus infinity, that is to say, it is not defined at zero.[3] As Figure 6.1 illustrates, the closer the values of x get to zero (from either side along the horizontal axis), the quicker $f(x)$ escapes towards (plus or minus) infinity, accordingly. This is expressed compactly as $\lim_{x \to 0} 1/x = \infty$ where zero is a point of impossibility.

Even though we can say that the ratio $1/x$ approaches infinity as x approaches zero, infinity is not a number, so the movement that 'erupts' in this way is away from the existing dimensionality towards another kind of space, towards a point radically different from all others in the space of original possibilities.[4]

Lacan identifies this phenomenon as one of the stages in the constitution of the subject: 'In so far as the primary signifier is pure non-sense, it becomes the bearer of the infinitization of the value of the subject, not open to all meanings, but abolishing them all' (Lacan, 1979, p.252). In this way, the subject is propelled into a radically different space, which transcends meaning.

Encounters with impossibility occur at those moments where continuity is lost. Mathematically, the question of continuity, as examined by Cantor, along the line of real numbers, was central to the development of set theory and, from there, to the topological understanding of space. Initially known as *analysis situs*, or the analysis of position, topology developed in the 19th century as a move away from a

Structures of the impossible 103

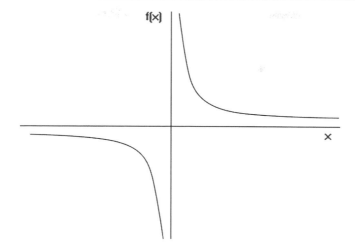

Figure 6.1 The multiplicative inverse function $f(x) = 1/x$.

mathematics concerned with precise solutions, allowing instead for the possibility of making qualitative predictions when quantitative ones were not an option (Totaro, 2008, p.382). Tucker and Bailey (1950) define topology as 'the mathematics of the possible' (p.18), its function being to settle the question of impossibility for other branches of mathematics. Importantly, topology can only illuminate whether solutions are possible, but not what they are or how to find them.

Topology focuses on the study of the geometrical properties that are not changed by continuous deformations, and applies to surfaces, knots and links (Stewart, 2013, pp.89–90). It is a rather general branch of mathematics, powerful in one sense and vague in another. It is not difficult to see in this particular position something of the foundations of its affinity with psychoanalysis.

In more specific terms, topology is a discipline developed around two key elements: the importance of continuity and the difficulty with defining discontinuity, of which holes are a radical manifestation. These two elements are also central to the development of psychoanalysis, which addresses dis/continuity and lack and impossibility at the level of human experience.

Paradoxically, although topology is concerned with space, topological objects can be described intrinsically, that is to say, without having to conceive of them as existing in some surrounding space in the way we ordinarily think of it (Stewart, 2013, p.98). Topological spaces are differentiated in terms of their structure, which in turn is defined by the presence or absence of twists and holes as manifestations of discontinuity. More on this in Chapter 7.

Of particular interest here are topological surfaces. One important feature of a topological surface is that it is 'in one piece' or path-connected, meaning that any two points on the surface in question can be joined by a continuous curve which lies entirely in that surface. In other words, it is possible to get from one point to another within the space in question by exclusively moving within that space. Movements across topological spaces are conceptualised as pathways or trails. Some such paths or trails close onto themselves, forming loops, and are known as cycles. For instance, every loop on a sphere can be continuously shrunk to a point (i.e. the sphere is simply connected and the loop is null homotopic). Also, every simple closed curve or loop drawn on a sphere operates as a boundary of the region it circles, separating the sphere into two distinct areas – inside and outside.[5] A cycle is a curve homeomorphic to a circle, so a closed pathway that does not intersect itself, even when it does not have the neat appearance of a circle. We can imagine stretching the loop, making it look tidy like a circle, without the need to cut it. Two cycles are of particular interest: those which designate a boundary and those which identify a hole. The main distinction between them is that those cycles that identify a hole are like a boundary without a territory; the manifold to which they could be the boundary is not there. Note that, in general terms, a two-dimensional manifold has the local topology of a plane, while a three-dimensional one has the local topology of ordinary, three-dimensional space as we know it (Weeks, 2002, p.40). So, in any 'small enough' subsection, topological spaces are no different from our most familiar notion of space.

Topologically, holes are structures which prevent an object from continuously being shrunk to a point. No matter how small the torus, it will never be reducible to something else, the hole that defines is configuration will always remain. Some pathways on a torus are reducible to a point, like on a sphere, and others not, as we will see in Chapter 7. The later type of loops are indicative of holes. This matters, because the presence or absence of holes and boundaries are central to the possibility of classification of topological spaces, which in turns makes it possible to know the structure of a space without 'seeing' it in its entirety.

6.2 Holes and the unconscious

Rigorous thinking about holes is rather sparse, yet intuition is not enough. It is perhaps easier to say what a hole is not, topologically, namely, not a cavity, such as a bowl, a cup, a bottle without its cap on. Vivier (2004) distinguishes two types of holes: cylinder-like and torus-like. The former is shaped like a pipe segment, the latter like the ring-like handle on a cup. A teapot without a lid has one of each types of holes: the pipe-like shape of the main body, with boundaries at the top and the end of the spout (cylinder), as well as the torus shape of the handle. A cylinder-type hole has two boundaries, while a torus-like hole has none (Vivier, 2004, p.142).

The classification of spaces rests largely on so-called invariants,[6] among which an important one is the genus, which denotes the maximum number of times a surface can be cut along disjoint closed curves or cycles, while remaining connected (i.e. in one piece). For a sphere, which has no boundary, the genus is zero. Cutting any slice off an apple disconnects it into two parts. Cutting a cylinder (think of a length of pipe) along the circumference separates it into two shorter cylinders, while cutting it along its length turns into a rectangle which can be laid flat. Both the sphere and the cylinder have genus zero. On the other hand, a torus – think of a bagel, which, like an apple, is also without a boundary – can be cut twice around its consistency (i.e. that which is not hole): the first cut makes it into an open C-shape and the second produces two smaller C-shapes. The torus has a genus one. In general terms, the genus corresponds to the number of (topological) holes in a surface.[7] We merely preface here the importance of the cut as analytic intervention, by recognising that the nature of a cut and the number of cuts have specific effects in modifying a particular spatial structure.

Spatially, holes are manifestations of radical discontinuity, of impossibility. If we think of the unconscious as a space, then the topological nature of this space has a lot to tell us about any structural impossibility at play and about ways to engage with navigating this.

Rosolato (1978) demarcates three stages in Freud's work, distinguishable in terms of the relation to the unknown that underpins them: first, the period of establishing signifying structures, starting with the interpretation of dreams; second, the work around the second topography and the death instinct; third, the focus on femininity and the relation to the mother as ultimately ungraspable reality (p.254). The sense of impossibility is present in Freud from the start, in the recognition that something remains ultimately out of reach in every production of the unconscious, what he calls the navel of the dream (Freud, 1900). Rosolato regards the navel as the vestigial mark of the link with the mother (1978, p.256), and calls it a 'blind hole' (p.257), a mark of discontinuity (p.258), a 'maternal slot' (p.259) that opens into what Freud called the *Unerkannt*, the unrecognised, the original Unknown (p.260). For Rosolato (1978), the signifier of the lack plays an essential role, organising the distinctions between the unrecognised, the knowable unknown and the unknowable unknown as absolute limit (p.262). This is one instance where the signifier operates as a navigation system for a space that cannot be otherwise mapped or traversed.

Following Lacan, Vandermersch (2009) designates the unconscious itself as a gap, a failure of production or of satisfaction which is not merely inaccessible but impossible. However, he does not conceive of it in purely spatial terms, but as a spatio-temporal gap (p.139).

Also, like Lacan, Bursztein (2017a, 2017b) conceptualises the interplay between the registers of experience (Real, Symbolic and Imaginary) as a spatial configuration where each dimension operates alone and in interaction with the others, to demarcate a landscape defined by holes. He maps each of these as part of the articulation of the relationship between desire, language and the body (recall the introduction to the Borromean knot in Chapter 3 and keep this in mind when we turn to elements of knot theory in Chapter 8).

These rather abstract formulations of impossibility are, at the level of experience, encountered daily in terms of lack and our oblique dance around the unspoken question of death. As Yalom (1980) puts it directly, death 'is a primordial source of anxiety and, as such, is the primary fount of psychopathology' (p.29). It

1	2	3	
5	6	7	4
9	10	11	8
13	14	15	12

Figure 6.2 Jeu de Tacquin.

is one manifestation or another of this existential predicament that brings anyone to the consulting room of the psychoanalyst. The wish to get rid of the lack is potent, yet it is lack itself which makes movement, and therefore life, possible. When it comes to the importance of incompleteness for the possibility of change, an illustration I often use in my lectures, and sometimes with patients, is the Jeu de Tacquin, invented in 1878 (see Figure 6.2). With its 16 positions of which only 15 are occupied at any one moment, the game affords some 10 trillion possible configurations, with just as many impossible ones (Kasner and Newman 1940, p.178). Had the game had 16 blocks and no gap, it would have been inert and of no interest.

The change from one possible configuration to another is, essentially, produced not by moving the numbered slates, as much as by 'moving the blank space itself through a specific path' (Kasner and Newman 1940, p.179). In other words, what is possible can be accessed by utilising the gap, without ever addressing it.

6.3 The Oedipus complex as prohibition veiling impossibility

As you may recall from Chapter 4, at the end of his adventure, the Square is caught up between prohibition and impossibility. He is punished for speaking of higher spatial realms but it is not the imprisonment that prevents him from saying what he wants to convey; it is the nature of radical incompatibility between spaces of differing dimensions that prevents him from ever bringing his extraordinary experience into the world he inhabits. This is the Square's plight.

Psychoanalytic understanding of human suffering has changed greatly over the last century. Freud's initial formulations rested on the notion of prohibition encapsulated in the Oedipus complex. We suffer because we cannot easily accept castration and because we desire what we are not supposed to have. Our drives are in conflict with the demands of civilisation, and this leads to a conflict between our desires and that which we take as deemed acceptable by others and our self-image formed through them. The most radical departure from this view, which remains prevalent in the IPA, comes, once more, from Lacan's later work. Moving beyond his initial reading of Freud on the structuring importance of each subject's response to the Oedipal predicament, Lacan makes it clear that the prohibition that castration is associated with is merely a veil that hides an impossibility. Castration, he clarified, pertains to the order of the symbolic, yet something of the lack encountered by the subject goes beyond symbolic capture – privation is real. At the core of experience there is 'a lack for which the symbol cannot compensate' (Lacan, 2014, p.136), 'an irreducible structural defect' (Greenshields, 2017, p.78), a paradox that Lacan saw as embodied in the structure of the torus, a spatial configuration which is complete but also has a hole.

As we saw in Chapter 3, the navel is, according to Rosolato (1978), a Freudian metaphor which points at a primordial, maternal unknown, one that cannot be symbolised, an unknown turned into a prohibition by the paternal metaphor. Against this, the unconscious amounts to a particular relation to the unknown and to death as unthinkable (p.267). In his view, Freud's introduction of the death drive amounts to an invitation to think of a limit, of a centre stripped of signifiers, a Nothing which is the constitutive source of the psychical apparatus, akin to the place zero holds among numbers. Remember that, in mathematics, the concept of the limit 'arises properly only in connection with infinite processes' (Kasner and Newman, 1940, p.235). As Greenshields (2017) underlines, Lacan turned Freud's metaphor into a topological Real (p.31), and tasked psychoanalysis with isolating this anti-conceptual hole without necessarily giving it a meaningful articulation (Lacan, 1979, pp.22–23). As we are going to see in Chapter 7, in his exploration of the hole as a structural feature of the torus, Lacan is careful to distinguish between nothing and void. The nothing pertains to the endless circling of the hole where some subjective patch (what Lacan calls the object-cause-of-desire) is placed by the subject, whilst the void concerns the tubular space circled by demand, in a way which is destined to fail, in order to sustain the pursuit of life through movement (Lacan, 2020, pp.369–386). Demands fail in such a way that desire can be sustained.

Satisfaction is not elusive merely because it could be found only in a forbidden space barred by a 'no entry' sign, but rather, because, where we expect to find it, there is nothing but a hole. We try to fill this hole with phantasy and love objects, we select something that might satisfy the drive, yet – without fail – it fails, everything fails. Complete satisfaction is not merely forbidden or out of reach, but fundamentally and structurally impossible. The failure is necessary, however, and it operates as an engine of search, tracing in this way the path of desire and therefore of life.

In a careful reading of Lacan's evolving views on the unconscious, Verhaeghe (2018) examines in detail the causal function of the original loss at the level of being (p.236) posited by Lacan. The loss in question begins with birth, whereby reproduction involves the loss of cell material from the formation of the embryo (Verhaeghe, 2018, pp.241–242, 251). This is followed by another loss, when the subject loses itself by entering language, precisely as a way of becoming a subject: 'when being makes its appearance at the level of language, it disappears under that language', leaving each of us 'with a basic lack as a condition for our becoming' (Verhaeghe, 2018, p.238). This is what Lacan calls the *manque-à-être*, sometimes translated as want-to-be or lack of being, as the underpinning of desire itself:

> Desire is a relation of being to lack. This lack is the lack of being properly speaking. It isn't the lack of this or that, but lack of being whereby the being exists. This lack is beyond anything which can represent it. It is only ever represented as a reflection on a veil.
>
> (Lacan, 1991, p.223)

In this sense, one is destined to repeat in relationships something about '[T]he original gap between life and death, between the body and the I, between the subject and the Other' (Verhaeghe, 2018, p.252), setting in motion a repetition in pursuit of a satisfaction that inevitably ends in failure. In some sense, the satisfaction – of life itself, if not of the subject – is in the pursuit of satisfaction itself.

The lack at stake is the mark of mortality. Topologically, this is the hole. Although it is usually Lacan's later work that one associates with topology, it is as early as Seminar II, where the citation above originates, that he clarifies this link. If knowledge encapsulated in the structure of topology is rejected, the possibility of immortality follows within the particular logic of the unconscious: 'patients who do not have a mouth or stomach will never die' (Lacan, 1991, p.237). 'What they have identified with is an image where every gap, every aspiration, every emptiness of desire is lacking [...]' (p.238). In less stark cases, one often encounters in the clinic stories of people turning away from their own desire, not showing up, playing dead in their own lives, as a strategy to keep themselves out of the circle of life, and therefore of mortality. At the same time, some intrinsic knowledge about the futility of this strategy pushes them to pursue speech as a way of inventing a compromise with being alive and therefore mortal.

The lack manifests as alienation both in relation to the body and in the realm of language, which each subject finds ready-made in the world inhabited by the body, and which each one enters rather than creates. However, the alienation in the language of the Other need not be the end of the story. Indeed, what analysis can offer, precisely through the use of language and its yonder, is the possibility of choice and change 'beyond the determination coming from the Other' (Verhaeghe, 2018, p.252), that is to say, the possibility of separation (Lacan, 1979, p.214). Lacan explores this theme using a Venn diagram between the subject and the Other, and between being and meaning, understood as the operation of intersection between sets, which are a particular representation of space. The intersection between the subject and the Other is itself a space 'where something is lost forever, that keeps operating as a force field' (Verhaeghe, 2018, p.243), with libido as an immaterial organ (*ibid.*). In this sense, the unconscious operates as an ever-failing border process, 'an opening, a gap, a crevice' (p.245). Its spatiality is recognised in this language, and topology is the discipline that can articulate something about the nature and structure of such a space. This is a space which can only be encountered as a sequence of repeated failures. Both the repetition and the failure point at impossibility, and the subject is constituted in relation to this 'hole' of the overlap with the other, where the function of the unconscious operates.

The movements in such a space, which is marked by discontinuity and impossibility, as well as the links between the signifiers that inhabit it, were seen by early Lacan as governed by mathematical rules that determined 'the possibilities of circulation and production of these signifiers' (Verhaeghe, 2018, p.247). In other words, it is language that makes some movement possible in relation to a space that cannot be accessed because it does not exist as such in the space inhabited by the body.

Lacan elaborates several topological conceptions on the unconscious, an essential one being the one put forward in *The position of the unconscious* (Lacan, 1966), where he introduces the unconscious as an edge, a rim that opens and closes at the same time, an in-between that 'does not have any ontological status whatsoever' (Verhaeghe, 2018, p.234). This particular description brings together notions of dis/continuity with the experience of repetition. Repetition itself is an inscription of discontinuity, so, topologically, the mark of a kind of movement circumscribing a hole, while the symptom itself is an iteration of the One which repeats itself, but cannot be erased. This impossibility can be understood in two ways: either the One is located in another dimension, therefore always on the 'outside' of the space of subjective being, yet present in it as a shadow, or it can be understood as a hole.

In his later work, Lacan moved away from language and the function of rules as defining of impossibility, to structural impossibility as embodied in topological spaces, making use of manifolds in order to demonstrate how the subject is formed through 'internal exclusions and external inclusions' (Blum and Secor, 2011, p.1030). Blum and Secor propose the term of 'psychotopology' in order to capture the ways in which 'material and psychic spaces are inseparable from one another' (Blum and Secor, 2011, p.1031). More specifically, they use topology to explore repetition in neurosis, with an illustration using Freud's Rat Man case (Freud, 1909), showing how structural relationships persist despite processes of transformation undergone by the psyche in an analysis, with the subject caught up in the topological complexity of their own psychic space. Indeed, to build upon their argument, one can understand transference in the same way, as what is at stake is not the localisation (in terms of time and place) of psychic registrations, but the relationship between these. Indeed, in Seminar XX, Lacan makes explicit his position regarding the 'strict equivalence between topology and structure' (Lacan, 1988, p.9).

Cléro (2002) declares holes as a constant of Lacanian topology, which engages with the orifices of the body as well as with psychic holes (p.85). This fundamental notion is at the heart of understanding the registers of experience proposed by Lacan, in their insolvable interplay: each registers impacts on the other two, whilst also failing in relation to them. Thus, the Borromean knot, which interlinks the Real, the Imaginary and the Symbolic, is essentially articulated around a hole, a fundamental lack between three heterogenous elements (Cléro, 2002, p.86).

Miller (1968) emphasises the relationship between structure and lack and examines the ways in which the invisible accommodates a structure which systematises the visible that hides it. This is very much a summary of the relationship between spatial dimensions. According to Miller, the lack that persists in the subject is the result of this very structure located in the invisible, so to say, and this alienation cannot be transcended in order to be understood. The impossibility of such transcendence and the absence of a metalanguage and of a meta-psychological position are all psychic expressions of the impossibility of movement from a lower to a higher spatial dimension from within the space of lower dimensionality.

6.4 Negations of impossibility

Much of psychotherapy and psychoanalysis follows a medical model, where certain manifestations are regarded as symptoms of particular pathologies, to be cured, so that the patient 'recovers' and returns to 'normal'. Radically different, Lacanian analysis does not regard anything as a predetermined symptom of any particular illness. Instead, the focus is on listening out for each person's mode of relating to the signifier, given that the unconscious itself is made of signifiers through which the dimension of internal Otherness specific to the unconscious is expressed. It is from this place that the questions relating to their very existence are formulated for each subject. As sharply put by Vanheule (2020), '[F]or humans, death is not just a biological matter: it is an existential event in terms of which we live' (p.181). Our starting point always seems to be that 'one's life is of value in terms of the Other's desire' (p.182). The answer each of us constructs in relation to the monumental question of how to be both alive and mortal can be understood in terms of some structural variations which are, in turn, determined by our very personal relationship to impossibility. Attention to these structural features is intrinsic to the Lacanian clinic. 'Far from being hidden underlying entities that are to be discovered in the depth of the human mind, clinical structures are nothing but patterns that we can detect at the surface of what people articulate' (Vanheule, 2020, p.201).

The Lacanian clinic distinguishes between three main structurally defined positions: neurosis, psychosis and perversion (see, e.g. Fink, 1997). The differentiation originates in the way each handles what Verhaeghe calls 'the original anxiety-provoking situation, that is, the confrontation with the first Other' (Verhaeghe, 1996, p.7), namely, the lacking mother and the associated threat of becoming the object of her enjoyment and of disappearing in it. What terrifies is not the discovery of one's own lack, but of her lack. In this reading, the father comes on as 'the second great Other, the Other of the Law that regulates enjoyment and pleasure' (Verhaeghe, 1996, p.5). This is the Oedipus configuration: our existence arises in relation to another, the m(O)other by which we remain defined and from which we need to separate in order to exist beyond biology. The fact that the other lacks gives us a place to strive for, invites us to respond to their desire, but also threatens our being with the danger of becoming nothing more than an object for them. The function of the father comes in to facilitate this separation, enough for the subject to emerge. Except, there is no guarantee that the operation succeeds. Furthermore, the prohibition of the paternal intervention aims to introduce merely a veil to the even greater threat to our subjective existence, that is to say, our inbuilt mortality.

In Lacan's words, a person expresses himself as the ego in *Verneinung* or denial (Lacan, 1951, p.11), as the essential function of the ego is a systematic refusal to acknowledge reality (p.12). His definition of psychic structures rests on the differentiation between three types of responses to the impossibility veiled by the threat of castration. Thus, in neurosis, the structuring solution is that of repression, which achieves a veiling effect. In perversion (not universally accepted as a distinct structure), the solution is that of denial, which amounts to a rejection of castration. Finally, in psychosis, the encounter with castration leads to foreclosure. One could argue that in psychosis the encounter with impossibility is the least mediated: as the prohibition of the Name of the Father[8] fails to inscribe itself, the subject fails to enter fully the Symbolic register, and impossibility is met with impossibility. Lacan opposes the notion of foreclosure to what Freud calls *Bejahung*, and which is defining of neurosis, as the mental process whereby attributes are attributed to an object (judgement of attribution), transforming a perception into a representation (Freud, 1925, p.236). In Lacanian terms, there is a radical difference between instances where this process is possible, and an inscription of the Name of the Father occurs, and those where perception exists but no such mental representation is created (i.e. in psychosis). Whereas in neurosis this particular signifier operates as an explanation for the mother's absence and jouissance, in psychosis, its absence leaves the jouissance of the (m)Other as a 'strange enigma' (Vanheule, 2020, p.189). Instead, in neurosis, the Name of the Father is what the subject/infant assumes that causes desire: 'The phallus is the name Lacan gives to this presumed cause: the phallus is a signifier that the speaking subject searches for in pursuit of that which causes desire' (Vanheule, 2020, p.186). For those not familiar with Lacan's terminology, this can be a rather tight and obscure fragment of text. Without going into too much detail, suffice it for now to spell out something about the concept of jouissance, perhaps the most Lacanian concept of all, which is central to engaging with the examination of clinical implications in Part IV. In essence, jouissance captures the satisfaction that reaches beyond the pleasure principle, through pleasure to pain. The subject's pursuit of it is unconscious, but unmistakeable. Lacan situates the signifier in the primitive rapport between knowledge and jouissance (Lacan, 2007).

Rosolato (1978) explores each of the specific responses in relation to the loss of the object – the mother, the breast, the phallus that would have existed in the intimate union with the mother and from which nothing remains (p.258). The original phantasy of turning towards the origin, that is to say, towards the mother's body, operates as an obturation or veil of the impossibility that marks the loss of something that never existed, namely, immortality, which was lost by dint of being born. Only living things die.

The best one can hope for, as a biological entity, supported as subject but also constrained by language, is not a signifier for what cannot be, but one that operates as a placeholder for the lack itself. This is what Lacan calls the phallus. Different relations to the signifier of the lack produce different structural outcomes. When the lack at stake can be symbolised, the hole can be 'filled' by various objects, as it happens in neurosis. The known and the unknown can be experienced as either relative or absolute. In the case of perversion, the unknown and the known coexist, without the contradiction at play being verbalised: the disavowal of the mother without a penis and the recognition of the same body are resolved through the fetish which cancels out the split. Finally, in the case of psychosis, where the symbolisation of the lack cannot occur, the signifier for the lack itself is lacking, and this is the case of foreclosure as a form of rejection of the impossible. All that can register somehow are the signifiers of the mother, which try – and often fail – to plug the breach.

The hole as marker of impossibility is universal, and defining of human experience, even though the way in which each subject engages with its structural impact varies. We now turn to an exploration of the main insights made available by topology to understanding possible ways of engagement with spaces defined by holes. These give some orientation with what is possible about the impossible.

Notes

1 Singularities are points which are not smooth, points where the nature of the object or space in question changes to something radically different compared to all other points.
2 See Verhaeghe (1996) for an overview.
3 A reminder here that the pairs of points $(x, f(x))$ that satisfy this relationship constitute the Cartesian coordinates of one point in the plane, and the curve $f(x)$ that depicts the function is composed of all the points where the relation specified by the function is satisfied.
4 This impossibility is resolved in the context of complex analysis, where a new kind of space is defined by adding to the complex plane (as depicted in Figure 3.2) a point at infinity; this is known as Riemann sphere. The discontinuity introduced elsewhere resolves the discontinuity around the origin illustrated in Figure 6.1.
5 This is explored in more detail in Chapter 8, where we consider the notion of what is known as a Jordan curve, namely, a continuous curve which divides the plane into two regions, an interior and an exterior (Crilly, 1999, p.10).
6 An invariant is a measure associated with topological space that does not change under continuous deformations of that space.
7 Although ordinarily we think of a tube as having a hole, topologically that is not the case.
8 Like many other concepts, this is a notion which evolves over the span of Lacan's work: whilst initially it refers to the function of introducing the law of the incest taboo, it later becomes essential to signification becoming fully possible for the subject, through the inscription – or lack thereof – of the original phallic signifier (Evans, 1996, p.119; Chemama and Vandermersch, 2009, pp.390–392).

References

Barrow, J.D. (1999). *Impossibility: The limits of science and the science of limits*. London: Vintage.
Blum, V. and Secor, A. (2011). Psychotopologies: Closing the circuit between psychic and material space. *Environment and Planning D: Society and Space*, 29, pp.1030–1047.
Bursztein, J-G. (2017a). *L'Inconscient, son espace-temps: Aristote, Lacan, Poincaré*. Paris: Hermann.

Bursztein, J-G. (2017b). *Subjective topology: A lexicon*. Paris: Hermann.
Chemama, R. and Vandermersch, B. (2009). *Dictionnaire de la psychanalyse*. Paris: Larousse.
Cléro, J-P. (2002). *Le vocabulaire de Lacan*. Paris: Ellipses.
Crilly, T. (1999). The emergence of topological dimension theory. In James, I.M. (ed.). *History of topology*. Oxford: Elsevier, pp.1–24.
Evans, D. (1996). *An introductory dictionary of Lacanian psychoanalysis*. London: Routledge.
Fink, B. (1997). *A clinical introduction to Lacanian psychoanalysis*. Harvard University Press: London.
Freud, S. (1900). The interpretation of dreams. SE4 & SE5.
Freud, S. (1909). Notes upon a case of obsessional neurosis. SE10, pp.153–318.
Freud, S. (1925). Negation. SE 19, pp.233–239.
Freud, S. (1937). Analysis terminable and interminable. SE23, pp.211–253.
Greenshields, W. (2017). *Writing the structures of the subject: Lacan and topology*. London: Palgrave Macmillan.
Kasner, E. and Newman, J. (1940). *Mathematics and the imagination*. New York: Dover Publications.
Lacan, J. (1951). Some reflections on the ego. *International Journal of Psychoanalysis*, 34, pp.11–17.
Lacan, J. (1966). Position of the unconscious. In Fink, B. (ed.). (2002). *Écrits*. London: WW Norton, pp.703–721.
Lacan, J. (1979[1973]). *The four fundamental concepts of psychoanalysis*. London: Penguin Books.
Lacan J. (1988[1975]). *The seminar. Book XX. Encore: On feminine sexuality, the limits of love and knowledge, 1972–1973*. London: WW Norton.
Lacan, J. (1991[1978]). *The seminar. Book II. The ego in Freud's theory and in the technique of psychoanalysis, 1954–1955*. London: WW Norton.
Lacan, J. (2007[1969–1970]). *The seminar. Book XVII. The other side of psychoanalysis*. London: WW Norton.
Lacan, J. (2014 [2004]). The seminar. Book X. *Anxiety, 1962–1963*. Cambridge: Polity Press.
Lacan, J. (2020). *L'identification: Séminaire 1961–1962*. Éditions de l'Association Lacanienne Internationale. Paris: Publication hors commerce.
Laplanche, J. and Pontalis, J.B. (1988). *The language of psychoanalysis*. London: Karnac Books.
Marcus, S. (2014). *Singurătatea matematicianului: discurs Academia Română, București 2008*. Bucharest: Spandugino.
Miller, J-A. (1968). Action de la structure. *Cahier pour l'Analyse*, 9.
Rosolato, G. (1978). *La relation d'inconnu*. Paris: Galimard.
Seife, C. (2000). *Zero: The biography of a dangerous idea*. Chippenham: Souvenir Press.
Stewart, I. (2013). *Seventeen equations that changed the world*. London: Profile Books.
Totaro, B. (2008). Algebraic topology. In Gowers, T. (ed.). *The Princeton companion to mathematics*. Oxford: Princeton University Press.
Tucker, A.W. and Bailey, H. S. Jr. (1950). Topology. *Scientific American*, 182(1), pp.18–25.
Vandermersch, B. (2009). Littoral ou topologie du refoulement (suite). Comme rien n'est simple… *La revue lacanienne*, 2(4), pp.115–119.

Vanheule, S. (2020). On a question prior to any possible treatment of psychosis. In Hook, D., Neill, C.and Vanheule, S. (eds.). *Reading Lacan's Écrits: From 'The Freudian thing' to 'Remarks on Daniel Lagache'*. London: Routledge, pp.163–205.

Verhaeghe P. (1996). The riddle of castration anxiety: Lacan beyond Freud. In *The Letter. Lacanian Perspectives on Psychoanalysis*, 6, Spring, pp.44–54.

Verhaeghe, P. (2018). Position of the unconscious. In Vanheule, S., Hook, D. and Callum, N. (eds.). *Reading Lacan's Écrits: From 'Signification of the phallus' to 'Metaphor of the subject'*. London: Routledge, pp.224–258.

Vivier, L. (2004). *La topologie: L'infini matrisé*. Paris: Le Pommier.

Weeks, J.R. (2002). *The shape of space*. 2nd ed. London: CRC Press.

Yalom, I.D. (1980). *Existential psychotherapy*. Basic Books.

Chapter 7

The unconscious as topological space

In his intricate and illuminating exploration of the questions that topology can answer for psychoanalysis, Greenshields (2017) singles out its importance 'as a writing of discomfortingly paradoxical and unfamiliar spaces and dynamics' (p.6), difficulties and paradoxes that it is 'called upon to present rather than resolve' (p.8).

Lacan is radical in his recognition of topology as structure itself, and not as a mere 'cartographic illustration of structure' (Greenshields, 2017, p.62). According to Lacan, topology is 'not "designed to guide us" in structure. It is this structure […]' (Lacan, 1972, p.483). In his wake, Bursztein (2017) singles out topology as that conception of space which resonates with subjective space, which is one of lack (p.13). What topology can offer uniquely to psychoanalysis is the way in which it can dramatise paradox, in the sense of *mise-en-scène* (Hughes, 2013, p.58).

Given that we are now considering the unconscious as a topological space, it is opportune to spell out what a topological space is mathematically, namely, 'a set of arbitrary elements (called points of the space) in which a concept of continuity is defined' (Alexandroff, 1961, p.8). This concept of continuity is based on the existence of local or neighbourhood relations which are preserved under continuous transformations (mapping). Mathematically, a transformation is a function that establishes a particular correspondence between all the elements of the set or space undergoing this change and their modified state, a way of establishing 'a correspondence between points according to some rule or law' (Barr, 1964, p.185). Of particular interest are structure-preserving, homeomorphic transformations, which are one-to-one and reversible. What psychoanalysis seeks from topology is the means to examine the correspondence and possible transformations between relationships, and this is what topology is best equipped to deliver.

If we can grasp something about enquiring into the relationship between spaces, then we are better placed to understand something about the spatial nature of the unconscious and therefore to intervene in such a way as to facilitate possible transformations, in full recognition of what remains intrinsically impossible.

After an introduction to the typology of two-dimensional manifolds in general, all of which feature in Lacanian and post-Lacanian thought, we return to the specific notion of topological hole and its relevance to the unconscious and to lack, and conclude by returning to the very idea of dimensionality which underpins the entire

DOI: 10.4324/9781003479284-10

exploration of the unconscious as space. The elements of topology introduced here enable a rigorous examination and re-framing of common misconceptions about internal/external aspects of experience and offer the grounds for a considered formulation of impossibility itself.

If we conceive of the unconscious as a topological space structured by discontinuity and grasp how some of the impossibility emerges as a consequence of dimensional impossibility, we are better placed to understand on new terms the relationship between body and mind and to intervene clinically in ways that engender changes, working with and around and in particular in relation to the subjective experience of impossibility.

7.1 Topology – a primer

The first definition of topology is attributed to Benedict Listing, a German mathematician of the 1800s who described it as the study of qualitative relationships of space (Charraud, 2004, p.140).

While the name of this new branch of mathematics was coined as early as 1847 (Epple, 1999, p.304), most of its body of work was developed only at the start of the 20th century. Part of its appeal was in that topology offered a view of infinity as tamed (Vivier, 2004, p.10), as it elaborated a mathematical structure that circumscribes infinity as a limit.

Topology is the study of those properties of spaces which remain unchanged under continuous transformations which can also be reversed continuously (i.e. topological equivalence or homeomorphisms) (Arnold, 2011, pp.122–123). Many of the concepts of topology were developed as generalisations of the properties of real numbers we explored in Chapter 3. Several branches of topology took shape, but the main focus here will be on what is known as algebraic topology, which concerns itself with the global rather than the local properties of a space. Local properties are those observable in small regions of the space in question, whereas global ones require consideration of that space as a whole (Weeks, 2002, p.39).

We focus here on two-dimensional spaces, or surfaces, as this is the type of space which supports the spatially grounded elaborations of both Freud and Lacan, as we have seen in Chapter 2. Incidentally, these are also the types of topological spaces about which most is studied and known mathematically.

A surface, in most general terms, is a geometrical shape that resembles a deformed plane. The most common understanding refers to boundaries of solid objects in ordinary three-dimensional Euclidean space (e.g. a sphere is the two-dimensional boundary of a solid ball). A surface is a two-dimensional space; that is to say, a moving point on a surface may travel in two radically different directions (it has two degrees of freedom). Locally, any surface is very much like a plane, in that every point has a disk-like or half-disk-like neighbourhood, depending on whether the space has a boundary or not. A surface is not porous; it is not a mere collection of points clustered together like bits of a comet. Instead, it is characterised by compactness.

Topological surfaces are classified as belonging to a surprisingly small number of equivalence classes. In other words, despite their vast complexity, each topological surface can be homeomorphic to one of a small number of standard ones (Zeeman, 1966, p.9).

Some surfaces are open, that is to say, they have a boundary, while others are closed. As far as the classification of closed surfaces is concerned, the key result is given by what is known as the classification theorem:[1]

> 'Any closed surface is homeomorphic either to the sphere, or to the sphere with a finite number of handles added, or to the sphere with a finite number of disks removed and replaced by Möbius strips. No two of these surfaces are homeomorphic.
>
> (Armstrong, 1983, p.18)

A closed surface is one which is compact, connected and has no boundary. Intuitively, this is a space which locally looks like the plane, that is to say, that each point on such a surface has a neighbourhood homeomorphic to the plane (Armstrong, 1983, p.149). If we were to draw a small disk around each point in this space, none of it would hang our over the edge, as disk *B* does in the space with boundary **S** in Figure 7.1. Rather, they would all have a contained neighbourhood, like *I*.

Before we examine the basic types of closed two-dimensional manifolds, we consider first an open surface which appears in a number of psychoanalytic formulations, namely, the Möbius band.

7.1.1 The Möbius band

One of the characters in *Flatterland*, the sequel to *Flatland*, is a cow called Moobius, who keeps her milk in a Klein bottle, is 'one side of beef' and has a resident (musical) band in and on it, the loss of which would be 'terribly orientating' (Stewart, 2002, p.97). This cow, whose description is rich in topological

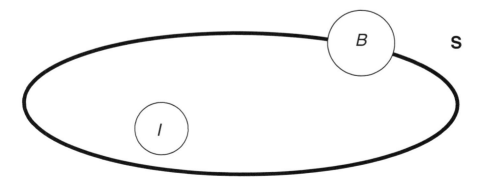

Figure 7.1 Open space boundary.

signification, introduces a crucial distinction between being locally two-sided and globally one-sided. This is not a partition whereby the two types of sidedness can be separated anatomically and placed next to each other. Rather, the one-sidedness of the whole manifests as two-sidedness of the part. What makes this possible is the twist in the 'tail' (Stewart, 2002, p.102). Let us learn to decipher something about this special cow and the world it inhabits.

The Möbius band, named after August Möbius, a contemporary of Listing's (who gave topology its first definition and who is also credited with the discovery of the same spatial structure a few weeks before Möbius), is a particular kind of space, a surface which has one side and one border, although it appears to have two of each. Locally, it has two sides at any one point, but not globally. It is non-orientable: an analogue clock, travelling alongside it all the way, would arrive at the starting point upside down compared to its orientation when it departed. This is one of several possible one-sided surfaces, the simplest version of which can be constructed by twisting once a strip of paper and then joining its ends. Depending on the direction of the twist, the band can be right-handed or left-handed. The two are equivalent (i.e. can be turned into one another) only in four dimensions (Wells, 1991, p.152).

The boundary of the band depicted in Figure 2.4 is a single circle. This boundary is constituted by the collection of all the points that have a half-disk-like neighbourhood. For each of the points in the interior of the band, we can draw around them an arbitrarily small circle. The disk that is delineated by this circle comprises points that are all on the band. However, for the points that are on the edge of this space, we can only demarcate a half-disk around each point; the other half would be 'hanging in the air', outside the surface of the band, as illustrated in Figure 7.1.

The less familiar fundamental polygon representation (see Figure 2.3) is important in order to get away from the intuition of depth that three-dimensional representations can evoke (Lacan, 2020, p.209). As Kasner and Newman point out, 'our intuitive notions about space almost invariably led us astray' (Kasner and Newman, 1940, p.183). More importantly, this representation, as opposed to the common picture of the twisted ribbon in Figure 2.4, completely determines the surface with regard to its intrinsic topological properties, away from any complications that arise from embedding this in three-dimensional space. As we will see, embedding creates problematic misperceptions of the spatial structures at play. In this representation, we can see that the Möbius band has two distinct, non-equivalent vertices, A and B, located in diagonally opposed corners, as Figure 7.2 illustrates.

Thus, the Möbius band is a surface with points on one pair of opposing edges identified with each other, one by one, as follows: A with A* and B with B*. The construction can be thought of as being obtained by zipping up the edges labelled *a*, in the direction of the arrows. Formally, the Möbius band can be written as $aba^{-1}b$, with the boundary composed of *a* and *b*, which constitute a single closed curve, topologically a circle ($a-1$ indicates the same edge, with one twist). This is the purely symbolic representation which moves away from the imaginary pitfalls of Figure 2.4.

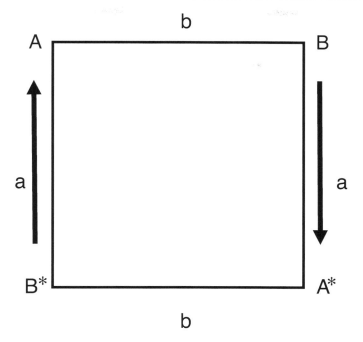

Figure 7.2 Möbius band in fundamental polygon representation.

The Möbius band is a one-sided surface, and is therefore non-orientable. Orientability is an intrinsic property of the surface, regardless of representation, whereas one-sidedness depends on the embedding of the surface in three-dimensional space (Seifert and Threlfall, 1980, p.9). This aspect is essential, in analytic terms, as it captures the inescapable spatial misperceptions engendered by living as a body in three-dimensional space. Orientability refers to the possibility of distinguishing left from right and up from down. We first encountered this in Chapter 4, in the description of the Penrose triangle and its 'verizontal' and 'inside-outerior'. This is a global rather than a local property, that is to say, orientable surfaces are those where such a consistent orientation can be maintained throughout their entirety, not only in arbitrarily small neighbourhoods. Any two-sided surface in space is orientable (e.g. disk, cylinder, sphere, torus).

One of the earliest topological objects studied, and perhaps one of the most familiar manifolds, with its global one-sidedness and local two-sidedness, the Möbius band is well suited to account for the moment to moment experience of inside-outside, internal-external, as well as for the overall sense of everything being both internal and somewhere else, out-there, at the same time. Lacan used the Möbius band as structure of the division between the subject and their utterances. Nasio links this to the question central to psychoanalysis: how is it that we can change ourselves through the mere act of speaking? (Nasio, 2010, p.16). What he singles out as a property of interest is that certain cuts to the band change the nature of the boundary and of the band itself. Following Lacan, he identifies speech with such a cut which splits the subject in two – while representing it, the signifier makes the subject disappear (Nasio, 2010, p.18). Likewise, Greenshields examines the Möbius band as the topology of 'the impossibility of the signifier that represents the subject coinciding with itself' and thus of the division of the subject (Greenshields, 2017, p.47). The twist that defines the structure of this space, but which has no particular location, is present everywhere through its impact, which makes it akin to the point at infinity that structures the cross-cap, of which the Möbius band is a part, as we are going to see.

7.1.2 The torus

Informally, the torus can be thought of as the surface of a ring doughnut,[2] as illustrated in Figure 7.3.

Formally, the torus is a closed surface, compact (i.e. it can be constructed by identifying the edges of a polygon) and connected (i.e. any two points on it can be joined by a path), with a hole. Being a closed space, the torus has no boundary. Given that a boundary is a set of boundary points, that is to say, of points with a half-disk-like neighbourhood, technically, the boundary for the torus is an empty set. Every point on it has a disk-like neighbourhood, meaning that we can draw a little circle around each point on the torus, and all the points within that circle would be still on the torus. It is equivalent to a sphere with a handle, and can also be described as the product of two circles – $S^1 x S^1 = T_1$, formally – or a circle of circles (Weeks, 2002, p.84). An arbitrary point on this surface can be described by a pair of coordinates,

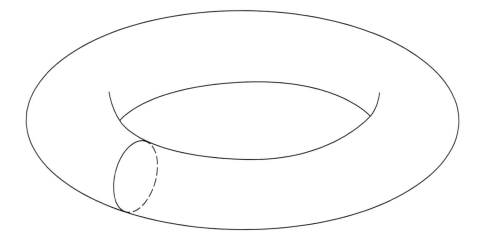

Figure 7.3 Embedded torus.

each of them being a point on one of the two circles (Mendelson, 1962, p.191). Each of the circles that are constitutive of the torus circumscribes a hole. As Greenshields (2017) stresses: '[T]hese holes/circuits are not the secondary features of the toric structure; the relationship between them is this structure' (p.67). Hilton (2008) goes even further by clarifying that, formally, 'the torus is itself a two-dimensional hole and any given point constitutes a zero-dimensional hole' (p.282).

In fundamental polygon representation, the torus is a surface where opposite edges are identified to each other in pairs, as in Figure 7.4. To construct it, one first zips up one pair of sides (e.g. those labelled a, in the direction of zipping indicated by the arrows), then the other pair (labelled b). Formally, this can be written as $aba^{-1}b^{-1}$ (e.g. moving counter-clockwise all the way around, starting from the top left corner).

The flat polygon and the doughnut surface are both representations of the torus. They have the same global topology, but their local geometries are different (Weeks, 2002, p.40). In other words, the flat representation is homogenous, while the doughnut is not. Counter-intuitively, the flat representation is a closed surface, All the vertices of this rectangle are one single point, A. If one deflates the torus

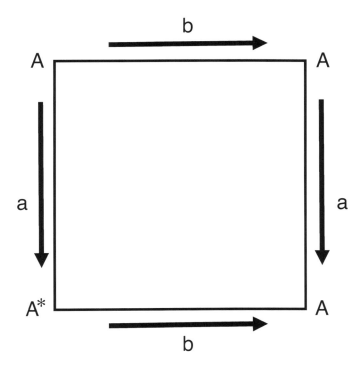

Figure 7.4 Torus in fundamental polygon representation.

and cuts it open into a rectangle, then rejoins one of the opposing pair of edges with a twist, one obtains a Möbius band. The converse operation is also possible. The torus does not contain a Möbius band, but it can be cut in such a way that the end result, a two-sided ribbon, is the same as the result of cutting a Möbius band along its middle. This means that it is possible to move between a torus and a Möbius band, by cutting and then stitching again, but differently.

Despite its appearance in the embedded version in Figure 7.3, the torus itself does not have a hole. Rather, what we observe as a hole is a consequence of embedding the surface into three-dimensional space. A 'proper' hole could be produced by cutting out a little disk on the surface of the torus, which would introduce a boundary and disconnectivity. This would be a proper hole in the fundamental polygon representation, but would appear like a cuff in the embedded version.

Movement between points on the embedded torus through a path that goes 'across the hole' entails access into another dimension, outside of the space itself. It is what for Lacan is the move outside of realm of sense, into nonsense, what he calls the realm of the *parlêtre*, that of the living body enjoying. What is more, the human body itself is homeomorphic to a torus, with the hole being that of the digestive system, from mouth to anus. The experience of discontinuity and impossibility in being occurs both at the level of the body and in language.

Lacan used the torus to examine repetition and the structure of the interplay between desire and demand (see Figure 7.5). Demand follows the repetitive circuit d and, through its repetition, inscribes the path of desire D, circling the hole where the object of desire is located by the subject. The pathway $d + D$ is what Lacan called the 'interior eight', best understood as a folded eight, with the upper part turned over the lower part and the two 'halves' sharing only one common point (see Nasio, 2010, p.16 and p.82). In terms of the symbolic representation, the torus is $T = D \times d$.

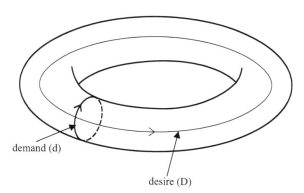

Figure 7.5 Demand and desire on the torus.

The fact that any point on the torus is defined by one coordinate on D and one on d indicates the inescapable, structural link between demand and desire at any one point in this space.

These circuits of demand and desire link to the distinction Lacan made between the nothing and the void, as we saw in Chapter 6. The same pathway that unifies demand and desire is also the one tracing a cut with transformative power in the cross-cap, as it is the one that separates the Möbius band from the disk, as we will see shortly. It is therefore recognisable as the edge of the Möbius band, namely, a circle that does not lie flat on a surface (the way the boundary of the disk does), but 'curls' above itself once in three-dimensional space. This edge joins the inside with the outside in an inescapable continuum.

7.1.3 The Klein bottle

Starting from the fundamental polygon representation (Figure 7.6), the Klein bottle can be constructed as a surface with the opposite edges zipped up, like for the torus – except that one pair is zipped up with a twist, as it was in the case of the Möbius band. In other words, this is a torus with a twist, and is also a closed surface, one without a boundary.

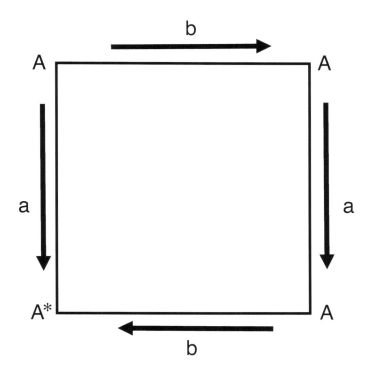

Figure 7.6 Klein bottle in fundamental polygon representation.

Another way to understand this topological surface is as a composite of two Möbius bands,[3] which makes the Klein bottle into a 'one-sided tube' (Seifert and Threlfall, 1980, p.13).

If we cut the fundamental polygon vertically in three strips (see Figure 7.7), we obtain one ready-made Möbius band, which is the middle strip, and another Möbius band which is obtained by joining up the other two strips along the side labelled c.

It is, like the Möbius bands, one-sided, but, unlike them, it has no edge, as the edges of the two bands are 'zipped up' to each other. Being one-sided, like the Möbius band, the Klein bottle is also non-orientable.

A Klein bottle can only be immersed rather than embedded in three-dimensional space, as it is not possible to construct it in three dimensions without self-intersection. However, this is possible in four dimensions, where the apparent absence of an inside vanishes. The usual representation of the three-dimensional version appears as self-intersecting along a circle, yet each of the points on that circle represents two different points on the bottle. It is important to remember that these points of self-intersection are not in the nature of the surfaces in question, but merely the consequence of forcing this surface into three-dimensional space (Figure 7.8)

As the representation in Figure 7.8 illustrates, the inside and outside are in a continuum with each other, whilst the surface is closed; it has no boundary. Hughes links this spatial structure to the uncanny, which 'brings us to a place where we have never been but which upon entering it, seems very familiar to us' (Hughes, 2013, p.74). Notice, yet again, that this conflicting perception is a consequence of our embodiment in three-dimensional space.

Lacan used the Klein bottle to elaborate his thinking about the continuum between psychical reality, material reality and the Real, in Seminar XII, of 1964–1965 (Lacan, 1965), but the same spatial structure also underpins his earlier concept of extimacy, the intimate exteriority of the Thing (Lacan, 1992, p.139), as well as constituting the structure of the space between signifiers (Nasio, 2010, p.13), conceived of as a set of elements which can only acquire consistency by lacking one element which, located on the outside of the set, becomes its boundary (Nasio, 2010, p.18). This already hints at the projective plane, to which we turn next.

7.1.4 The projective plane/cross-cap

Starting again from the fundamental polygon representation, the projective plane is constructed as a surface where each pair of opposed edges are zipped together, each pair with one twist (Figure 7.9). Like the Klein bottle, this is also a surface which cannot be embedded in three-dimensional space. It is the least easy to visualise two-dimensional manifold, as the self-intersection such embedding implies cannot be constructed in three-dimensional space at all. Both the Klein bottle and the cross-cap are four-dimensional objects represented inaccurately in the two-dimensional space of the page – or on Lacan's blackboard (Hughes, 2013, p.85) – as projections of apparently self-intersecting three-dimensional objects. In this sense, it is unhelpful to have a multitude of 'illustrations' of the projective plane,

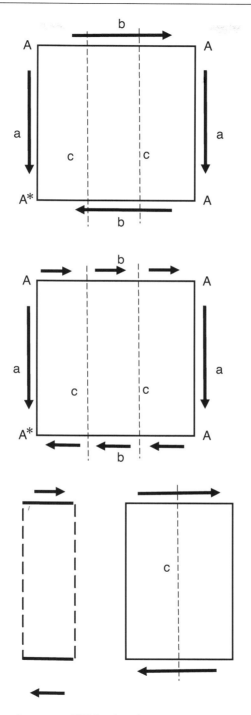

Figure 7.7 Klein bottle as two Möbius bands.

126 The unconscious as domain of impossibility

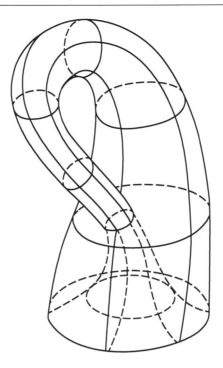

Figure 7.8 Klein bottle immersed in three-dimensional space.

as these are two-dimensional representations of something that cannot actually be represented accurately even in three dimensions. In four-dimensional space the self-intersection that prevents the representation problem is no longer there (Armstrong, 1983, p.17). In other words, this apparent self-intersection is not an intrinsic topological property of the projective plane, also called a closed Möbius band (Seifert and Treelfell, 1980, p.10).

The mock representations found in print are purely of an imaginary kind, devoid of symbolic value. Interestingly, the full difficulty of drawing (i.e. producing a two-dimensional representation) the projective plane was recognised only one century after the notion of such a space was conceived at the level of the Symbolic (Nasio, 2010, p.37).

Topology, as a discipline, is not restricted to the study of surfaces that can be embedded in three-dimensional space, and it is only the Symbolic that makes it possible to go far beyond what can be visualised, with proper understanding becoming available only by moving beyond the Imaginary.

The cross-cap is a homogenous surface, constructed from a punctured sphere, with antipodal points identified, that is to say, 'zipped' together in pairs, as illustrated in Figure 7.9, with the purely Symbolic form read as *abab*.

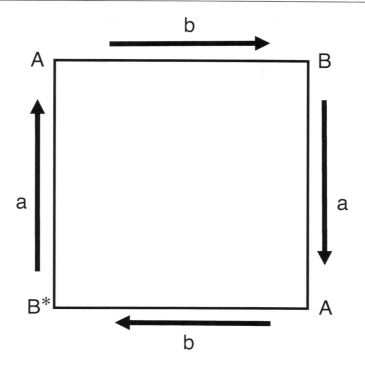

Figure 7.9 Projective plane in fundamental polygon representation.

Topologically, the surface sometimes referred to as the cross-cap, and which Lacan made reference to[4], is part of a representation of the projective plane, constituted by a plane and a point or line at infinity. This derives from the Euclidean notion of the plane, where any two lines intersect, except for parallel lines, which only meet at infinity. The addition of this point at infinity makes it so that all lines in the plane intersect, including parallel ones. It is, mathematically, a way of doing away with the impossibility. It is, I would argue, this point at infinity that is the point around which the cross-cap is organised, and this is also the place of the phallus in Lacanian thought. This 'hole' in the structure of the surface is where all lines meet, except they don't. It is a point of singularity. As Vandermersch explains, this point of exception is the one that gives sense to all others by cancelling itself out, safeguarding thus the possibility of nonsense (Vandermersch, 2008, p.8).

So while the lines of the Euclidean plane do not contain the point at infinity where they are presumed to meet, but rather they are all infinite themselves, the projective line contains this point at infinity. Furthermore, while Euclidean lines extend infinitely at both 'ends', every projective line is closed, like a circle joined up at this very point at infinity; in other words, the projective line is homeomorphic to a circle (Nasio, 2010, pp.33–35). The self-intersecting line of the cross-cap is this very line or point which constitutes the structuring hole, which is not a quirk of representation in limiting three-dimensional space, but a structural aspect.

The name of cross-cap comes from its resemblance with a bishop's mitre. Its local topology is that of a sphere (Weeks, 2002, p.59), but its global topology is radically different. The sphere with a disk removed is itself homeomorphic to a disk, whose edge is a circle. As the edge of a Möbius band is also a circle, the two boundaries can be 'zipped up' together, as a one-to-one correspondence can be established between the two boundaries. So, while it is impossible to fully visualise this topological structure, the cross-cap can be thought about as the combination between a disk and a Möbius band. One way of looking at the Möbius band here is as an edge built around a hole, with the composite structure amounting to what Lacan described as a certain way to organise a hole (Lacan, 2020, p.422). This becomes clear if we decompose the cross-cap into a sphere with a hole (a disk) and a Möbius band. The one-sided twisted band tries to cover this hole, but cannot restore the fullness of the sphere. Instead, it marks the edge of a persistent impossibility.[5]

This can be illustrated by cutting the fundamental polygon in Figure 7.9 into strips, then reorganising these as illustrated in Figure 7.10: zipping along edge **a** in the middle region produces a Möbius band. Joining together the other two rectangles, which are homeomorphic to two half-disks, produces the disk.

The boundary between the two edges is a circle which links two heterogeneous surfaces: a one-sided Möbius band and a two-sided disk (from zipping the remaining two sections along edge **b**, in the direction of the arrows). This composite of two open surfaces is a closed surface, without a boundary. As it contains a Möbius band (in the sense that a part of its surface is homeomorphic to one), the projective plane is a non-orientable surface.[6]

Reducing the Möbius band to the cut that engenders it and the punctured sphere to a point, the structure of the subject is in essence this: a line without points, together with a point outside this line (Charraud, 2004, p.147).

It is important to consider another approach to the projective plane, one not used by Lacan but nevertheless relevant to understanding the essence of this space. I am referring to the geometrical description of the projective plane, whereby each infinite line through the origin of the Euclidean three-dimensional space corresponds to a point of the projective plane, a projective point (The Open University, 2016, p.47). If we imagine a sphere centred on the origin, then each such line 'pierces' the sphere at two points, making each projective point equivalent to a pair of antipodal points together with the origin. This can be simplified as the points on one hemisphere together with the origin, which can be further simplified in the representation of a disk with antipodal boundary points identified. This version, as illustrated in Figure 7.11, was originally produced by Franz Klein, who also introduced the Klein bottle (Volkert, 2013). It is visible here how antipodal points are linked in pairs, defining a line that passes through a point that is not there, P. The edge of this dance around the hole is the Möbius band, as we have seen.

The steps of this transformation are well explained in Nasio (2010). Without reproducing his exposition here, suffice it to say that, in the move from the line to the

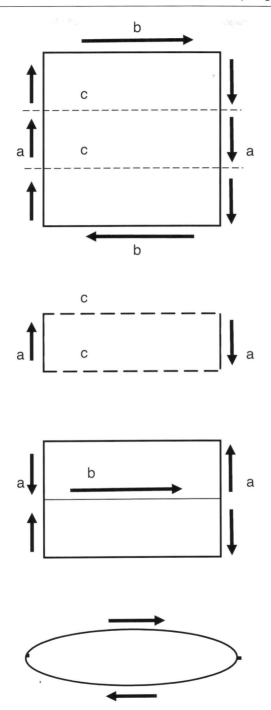

Figure 7.10 Decomposing the projective plane.

130 The unconscious as domain of impossibility

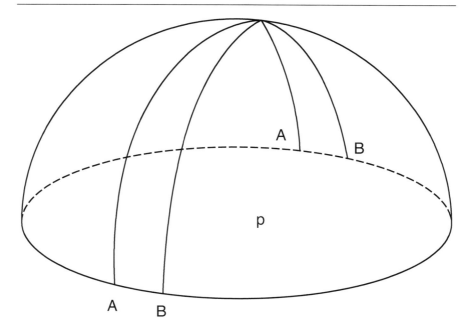

Figure 7.11 Projective plane alternative representation.

point, the bijection is lost, there is no longer a one-to-one correspondence between the two representations, as a consequence of the loss in dimensionality entailed in collapsing a line (two-dimensional) to a point (one-dimensional). This makes clear the asymmetric nature of movement between spaces of different dimensionalities, leading to the kinds of repetitions, overlaps, paradoxes and gaps that analysis aims to address clinically.

Lacan used the cross-cap as structure of the relation between the subject and their object, and therefore of phantasy (Nasio, 2010, p.14). The question this is addressing, in Nasio's words, is the following: 'how can we understand that what we call phantasy is not an image internal to the psychic economy of the subject, but an apparatus, a structural frame extended into reality, confounding itself with it?' (p.20, my translation). The cross-cap embodies the continuity between internal and external, as well as Lacan's formulation of phantasy as the relation between the subject and its object (which he writes as $\$ \lozenge a$). The subject is the one-sided Möbius, while the disk constitutes its object. As a topological surface, the disk can be continuously reduced to a point. Nasio (2010) reads this as the point at infinity to which every point on the boundary of the Möbius band is connected, thus being pulled towards the fourth dimension (p.98). In four dimensions, no self-intersection or impossibility would emerge. In Lacanian terms, the object of desire is an extrinsic vanishing point, presupposed mathematically, so as to close a space which is fundamentally impossible to close in its own dimensionality, that is to say, in the space inhabited by the body. This particular heterogeneity is particular to the cross-cap, which captures uniquely the radical alterity as well as the reversibility between the subject $ and the object *a* as mediated by phantasy (Vandermersch, 2008, p.8).

Thus, the cross-cap as deployed by Lacan gives materiality to three psychoanalytic concepts: the lack of distinction between inside and outside, the cut between the divided subject of the unconscious and object a (*obj a*) and the properties of *obj a*, the common element of all three being the phallus, whose signification appears in the cross-cap at the line of self-intersection. As far as the inside-outside divide is concerned, it is not a question of denying the existence of such a division, but of subverting it by engaging with the lack of separation between interior and exterior (Nasio, 2010, pp.78–79).

In terms of this particular spatiality, the subject can be understood as 'constructed' by screwing a Möbius band in the edges of a punctured sphere. Nasio (2010) goes as far as to read the deconstruction of the cross-cap into a Möbius band and a disk as a collage of a line without points (the band reduced to the line that describes its boundary) and a point without line (the disk reduced to a point) (p.99). He is not alone in stressing that the topology with which analysts operate is neither general nor algebraic topology as such, but rather a kind of artifice that dramatises the paradox of being (Nasio, 2010, p.23).

7.1.5 The cut

It is striking to know that, even before Freud had made his first publication in psychoanalysis, in mathematics it had already been established that every conceivable surface is a combination (formally, a connected sum)[7] of tori and/or projective planes (Weeks, 2002, p.75), while the sphere is the sum of zero tori and projective planes. The fundamental structure of space is very simple, yet its possible manifestations are endlessly complex. This simplicity can be captured most relevantly through the notion of the cut.

Poincaré (1912) emphasises the importance of the cut in topology: 'it is on the cut that everything rests' (p.488, my translation). With specific reference to the cross-cap, Nasio (2010) defines the cut as the line that separates and reunites two heterogeneous parts (p.72), and argues that it is the cut itself that reveals the structure of the surface at play. This is entirely aligned with Lacan, who goes as far as to argue that it is the cut that engenders the surface (Lacan, 2020).

In topology, a cut is the removal of certain arcs or segments from a set (Arnold, 2011, p.75). Lacan calls this a line without points (Lacan, 1972). Depending on the nature of the space where these arcs exist, the cut can lead to structural changes,[8] with local transformations producing a restructuring of the space as a whole. For instance, a space with a hole would support a number of pathways that would not be reducible to a point, but with the introduction of the cut, the nature of these pathways changes so that they become reducible. The impossible is transformed into possible and vice versa, as far as movement within the space in question is concerned. A cut can transform a space in such a way that new possibilities of movement arise.

If we also think of the space of language as a topological space, then certain cuts can free the links in the chain of signifiers and open it up in new places, creating an end where before there was a rigid and painful connection, opening a hetero closed path, or modifying the sequencing by allowing for new connections to form. More on this in Part IV.

7.2 Holes and the unconscious, revisited

For the purpose of this analysis, the term hole does not refer to a space where something used to be, but it has been removed or lost; rather, we keep closely to the topological understanding, which designates holes as a structural feature of space intrinsically linked to impossibility. Not only that there never was anything 'there', where the hole presents itself, but nothing of what is accessible can occupy and fill that space with any degree of success.

A poignant introduction to this particular kind of signification is offered in John Maxwell Coetzee's novel, *The Life and Times of Michael K*, where the life of one man is powerfully summed up thus: '*Always, when he tried to explain himself to himself, there remained a gap, a hole, a darkness before which his understanding baulked, into which it was useless to pour words. The words were eaten up, the gap remained. His was always a story with a hole in it: a wrong story, always wrong*' (pp.150–151).

Before turning to a more detailed consideration of the topological concept of hole, it may be helpful to take some orientation from Lacan in relation to the radical structuring role of the hole for the subject of the unconscious. In Seminar IX, referring to the constitutive object of phantasy, he clarifies this as follows: 'Every metaphor, including that of the symptom, tries to make this object emerge in its signification, but all the pullulation of meanings that it may engender never manages to staunch [seal] what is involved in this hole in terms of a central loss' (p.602) [my translation] (Lacan, 2020, p.467). This is precisely what Coetzee circumscribes in his novel.

Let us recall that in his pursuit of a spatial expression for his formulation of the structure of the psyche, Freud spoke of *topography*, although his language was intuitively topological. It was Lacan who made explicit the link between psychoanalysis and *topology*. While topography designates a precise description of a place, or a detailed graphic representation of the surface features of a place or object, where fixed positions are involved, topology refers to an entire area of mathematics concerned with spatial properties that are preserved under continuous deformations of objects (hence the name of rubber-sheet geometry). Both are concerned with surfaces, understood in the usual sense, as the outside layer of an object, but while topography is primarily attentive to measurement and fixing locations, topology is exclusively concerned with relative positions.

Although they play a central role in the classification of topological surfaces, and despite their structuring impact on any spaces, holes have no formal definition, and are not treated as a mathematical object. Instead, a hole is thought of as a particular form of discontinuity in a space that makes it possible to identify the nature or structure of that space as identical to or different from that of another (reference) space. In this sense, holes have no identity as such, but can help establish the identity of the spaces that they structure. This question can be approached intuitively by considering, for instance, whether a torus is the same with a sphere. If we imagine walking around on each of the two surfaces trailing a rope and returning back to the starting point, the existence of a hole can be settled by considering what happens when we return to the starting point and we pull the rope to us: if we can pull the entire loop back, while remaining on the surface, there is no hole; if it snags, there is. The former is true for walking around a giant apple; the latter holds for a walk on the surface of a giant bagel.

More formally, we can classify possible pathways on a surface (such as the one traced by the rope in our thought experiment) into those that can and those that cannot be reduced to a point. Whilst on the apple we can only have those that can be reduced to a point, on a bagel we have both types.

When it comes to topology, we see that the focus is on points of discontinuity and their interplay with continuity. We can thus understand that holes not as properties but as particulars, namely, as 'dependent particulars: they cannot exist alone; they need a host at the surface of which they find a place to be' (Casati and Varzi, 1994, p.6). A surface can exist without a hole, but not the other way around.

To use a much cited quasi-definition, 'A hole is there where something isn't' (Tucholsky, 1931). And there never was, psychoanalysis would add. The affinity of holes with Lacanian analysis is immediate, given that at the core of the work of analysis is analytic interpretation, which 'must be guided by the search for a truth effect conceived as a rupture' (Laurent, 2019, p.3). This is crucial, as it is only Lacanian analysis which makes the distinction between interpretation that produces meaning that can be understood and what can be called 'the truth effect of a fundamental void' (Laurent, 2019, p.5), linked as it is to the first absence, the lost object, the inaugural void (Laurent, 2019, p.6).

In this sense, interpretation operates as a kind of mapping from one void to another. Miller distinguishes between analytic interpretations which provide some elaboration and those which provoke perplexity, and cautions against 'too much meaning' (Miller, 1996, p.5), while Laurent proposes that the truth of the hole *introduces* the impossible (Laurent, 2019, p.7). Except, the impossible is already there, structurally, in the story and being of each and every one of us. Against the grain, interpretation, in speech or otherwise, indexes or designates the impossibility at stake. Equally, perplexity does not work without elaboration, in the same way that a hole does not exist in isolation from a space whose nature it defines and structures through its presence.

The essence of the fundamental lack inherent in being is succinctly yet precisely captured by Vanheule (2020), with reference to Lacan, when he recognises that 'the determination of human functioning by biology or by environmental factors is marked by a fundamental lack: for humans, fundamental questions concerning identity ("who am I?") and intentionality ("what do you want?") are never resolved' (p.199). Instead, we try to answer them through identifications. Within certain limits, subjects can give some shape to their own being: neurotics through the use of the Name of the Father as belief in the lawfulness of the world, psychotics through an invention of the dimension of the law not encapsulated by them in this un-inscribed signifier.

The lack-of-being that is at the centre of these unanswered or unanswerable questions is what conditions desire itself. For each subject, irrespective of psychic structure, there is an incompleteness at the level both of experience and of articulation. Yet, something is there, seen, but not what actually appears to be, as well as something that remains unperceived, like the dreamer in their own dream.

In his later work, Lacan (1974) coins the term *troumatisme*, translated by some as trouma, a blend of *trou* (hole) and *traumatisme* (trauma). In essence, language always fails and at best it chips away at the Real, where there is nothing to discover but a hole, in relation to which the unconscious is an invention that recognises and addresses this fundamental impossibility, that of life around death, what Yalom calls the ultimate concern (Yalom, 1980). Phantasy is a response to the encounter with this hole at the core of our experience of being.

7.3 Embedded or not – back to dimensions

The move away from Euclidean geometry disrupts established notions of dimensions. For instance, in Euclidean space, a cube is a three-dimensional object. In topological terms, it is just a two-dimensional surface, and is also equivalent to a sphere. Nevertheless, this surface is embedded in three-dimensional space. The space of immersion is not the same with the space of representation. For instance, Figure 7.3 representing a torus appears on a page, which is two-dimensional, which in turn sits in the three-dimensional space occupied by the book and the reader. The image is flat, and it depicts a three-dimensional appearance of a two-dimensional space.

The insertion of an object into a particular space can either keep the object unchanged (homeomorphic to itself) or modify it in such a way that it no longer remains equivalent to itself. The first scenario is that of embeddedness; the second is that of immersion. The intrinsic topology of a space is different from the topology of that space when embedded into another space (Stewart, 2002, p.105). Thus, the torus looks curved when embedded in three-dimensional space, but it is intrinsically flat. The doughnut shape and the tile with edges glued in the direction given by the arrows (Figure 7.4) are therefore equivalent but not identical. Embedding refers to a choice of how to place a surface in a space (Adams, 2000, p.78), whereas immersing refers to a local embedding, in the sense that locally there is a one-to-one correspondence, but not globally.

When the space in which the object is embedded has fewer dimensions that the object itself, we encounter something in the nature of an impossibility (e.g. a cross-cap cannot be embedded in three-dimensional space) or a false intersection (e.g. the appearance of a Klein bottle as embedded in three-dimensional space).

Both of these topological spaces, which are impossible or inaccurate in three-dimensional space, could be constructed correctly in four-dimensional space (Banchoff, 1996, p.201), except that is not a space that we can access as such.

If we recall the two conditions of homeomorphism (namely, that to each point in one space corresponds one and only one point in the other, and that to any two neighbouring points in one space corresponds to a unique pair of neighbouring points in the other), it is the first one, relating to bijection, that breaks down in the case of immersion. For instance, the self-intersection encountered when a Klein bottle or projective plane is immersed in three-dimensional space shows that loss of uniqueness: unique points on the fundamental polygon representation appeared merged along the line of self-intersection, where two or more (four, in the case of the cross-cap) distinct points in the former are replaced by only one point in the latter. In four-dimensional space, the self-intersection no longer occurs, and in both cases we can speak of these surfaces as embedded, that is to say, both the Klein bottle and the cross-cap would fit entirely and accurately in four-dimensional space. The apparent self-intersection which results from immersion in three-dimensional space makes the cross-cap appear like a closed two-sided surface, with an interior and an exterior, yet '[T]he cross-cap which interests us is certainly the one we can see, but to which we attribute the properties of the one we cannot see' (Nasio, 2010, p.71, my translation).

While embedding surfaces in three-dimensional space 'is justifiable as a natural intuitive aid' (Seifert and Threlfall, 1980, p.19), in order to explore higher-dimensional manifolds, freedom from surrounding space is necessary. Thus, the key notions of interest are those of neighbourhoods and their mapping. More importantly, the concept of neighbourhood makes it possible to define the most general concept of space, 'independent of vague intuition' (Seifert and Threlfall, 1980, p.20), and back into the abstract territory of set theory. It is at this level that we operate and intervene in our daily work in the clinic, that is to say, in the local spaces of signification neighbourhoods.

In his seminar on *Identification*, Lacan refers to what he can offer in his teaching as mere projections of a fourth dimension that cannot be represented (Lacan, 2020, p.268). It is interesting that he makes this particular comment at a point where he is expanding in detail his take on topology as useful to analysis, given that topology itself evolved historically as a response to difficulties with thinking of space simply in terms of dimensions. One can understand the imperfect correspondence between latent (four-dimensional) and manifest content (three-dimensional) in terms of the inherent loss of information that comes with moving from a higher to a lower spatial dimensionality.

Freud takes as given that 'latent mental states' [...] have abundant points of contact with conscious mental processes' (Freud, 1915, p.168). He explains that 'the different latent mental processes inferred by us enjoy a high degree of mutual independence, as though they had no connection with one another, and knew nothing of one another' (Freud, 1915, p.170), and proposes that 'internal objects are less unknowable than the external world' (Freud, 1915, p.171). Notice the spatial accuracy of his explanation, in terms of the mathematical elements examined so far, and keep in mind that the move from one dimension to another is not a continuous one, but a radical jump. One way of illustrating this is through the move from real to complex numbers, as we have seen in Chapter 3.

So how can a three-dimensional embodied being have a four-dimensional internal world?

One answer is that offered by Buddhism, where one does not have (contain) a mind (inside), but lives in one. The mind is both inside and outside; in other words, it pertains to another dimension. Something of this resonates with the recent proposed understanding of the mind as a receptor of waves originating outside the body (Christensen, 2017), in the same way that the eye is a receptor of light.

Indeed, it is because everything is there, in and all around us, rather than hidden, that it is so difficult to apprehend. The mind has a sky-like nature (Sogyal, 2002, p.12), penetrating in all directions and pervading everywhere (Sogyal, 2002, p.157); and the truth keeps interrupting us (Rinpoche, 2012, p.5). More specifically, in spatial terms, inside and outside 'are never separate or different; they are always the same' (Sogyal, 2002, p.49). In this sense, both the body and the mind operate as topological spaces, with the mind having one additional dimension as compared to the spatial world as perceived and inhabited by the body. Thus, spatially, being in a mind rather than having one can be understood in terms of the relationship between dimensions, given that 'life and death are in the mind, and nowhere else' (Sogyal, 2002, p.47).

Freud described his first encounters with the problem of loss of dimensionality in his early neurological work on speech disturbances, where he proposes that the representation of the body is only present as a projection on the nervous system, and favoured a move away from traditional spatial representation towards functional structuring (Freud, 1898), which in fact alludes to space in topological terms rather than Euclidean. As we have seen in Chapter 5, what appears as a sequence of spatial projections, distinct and similar, can be mathematically conceived of as a coherent entity of a higher dimension which cannot be incorporated directly in the space where these projections are encountered as partial, fragmented manifestations of the same whole.

Introducing the idea of a fourth dimension leans on the side of continuity; it tidies up things that appear as fragmented in three dimensions, repetitive and ill-connected, by making them link up in a way that can only be conceived of and not perceived as such. The idea of the unconscious itself was, similarly, an inference on Freud's part, an interpolation that could not be observed, but which, once posited, provided a link between conscious acts which would otherwise remain disconnected and unintelligible.

This exploration of the projection of higher dimension into lower ones says something about the difficulties with holding higher dimensions into lower ones. Perception can confound dimensions, but this can be adjusted or corrected to some extent. There is another aspect of the relationship between dimensions that escapes perception altogether, namely, the way in which higher dimensions hold within themselves lower ones. A higher dimension holds a lower dimension, but not like a container. Rather, the lower dimension is integrated into the structure of the higher space in a way that makes it inseparable from it. Higher dimensions become collapsed into lower ones, and reduced to shadows and repetitions. Each projection introduces an element of distortion.

Unlike higher dimensions, the fourth dimension has a sort of physical reality.[9] Traditionally, time has been regarded as the fourth dimension. A striking spatial formulation of this view, where time is expressed in terms of distance and the speed of light, is offered by Magee: 'Events not only in human history but throughout the whole history of the earth could be directly observed simultaneously by watchers on stars at different distances' (Magee, 2016, p.10).

The idea of a higher dimension of existence that cannot be embedded but can be engaged with on the level of speech is that of the infinity of infinities: 'one cannot trap or exhaust Ein Sof with mind or story or intention' (Eigen, 2012, p.82).

Lacan recognises the limitations of representation of higher-dimensional objects in lower dimensions. (Lacan, 2020, p.268) and acknowledges the problem of embeddedness, which is also subject to the anatomical constraint of the field of vision as essentially two-dimensional (Lacan, 2020, p.274). He poses the question of the structure of space in terms of what kind of surfaces are possible in it, and at the same time, he remains rather dismissive of the idea of a fourth dimension, especially in the context of his rereading of Schreber's memoirs and of Freud's analysis of this text in his 1955–1956 seminar on psychosis (Lacan, 1997). He regards Schreber's appeal to a fourth dimension as a common solution to a metaphysical dilemma: '[O]ne makes do with saying that somewhere there is a fourth dimension and a diagonal', the solution reached for when 'one has absolutely no idea how to reconcile two terms' (Lacan, 1997, p.68). He locates Schreber's experiences 'in what could be called a trans-space linked to the structure of the signifier and of meaning, a spatialisation prior to any possible dualization of the phenomenon of language' (Lacan, 1997, p.141), with 'space speaking as such' (Lacan, 1997, p.142). Yet something crucial emerges from his objection, as this is the first explicit reference to the idea of a subjective topology which he is to develop in his later work, a development which transformed the essence of the work we are all engaged in, whether we know it or not: ethically, psychoanalysis is guided by the search for 'pathways through spaces that give access to desire' (Burgoyne, 2018, p.17).

Before moving on to a closer examination of the clinical relevance of such pathways, we consider a particular formulation of spatial movement trajectories, in the form of knots.

Notes

1 This theorem is also owed to Möbius (Armstrong, 1983, p.18).
2 We are considering here only the one-holed torus, but *n*-holed tori are possible, with *n* taking the value of any integer from 1 upwards. These would also be considered equivalent to a sphere with *n* handles.
3 Although, strictly speaking, a Klein bottle contains a multitude of Möbius bands, as any strip, straight or wavy, which cuts across horizontally will have the opposed ends zipped with a twist (Weeks, 2002, p.51).
4 Strictly speaking, the surface Lacan refers to as the cross-cap is the equivalent of a cross-cap with a sphere (Mendelson, 1962, p.198). This, in turn, is a visible representation

of the projective plane, which is an abstract two-dimensional surface (Nasio, 2010, p.28). Lacan did not always respect the established meaning of the mathematical concepts he incorporated in his work. Thus, he referred to the Borromean link as the Borromean knot (but it is not a mathematical knot), and used the notion of cross-cap to refer to the projective plane. These two are not always clearly differentiated in some mathematical texts either, yet they are not the same thing.

5 Note that both the torus and the cross-cap start from a sphere, but modify it in important ways, producing what Lacan called l'asphere (Lacan, 1972, p.29).

6 Conversely, since the projective plane is represented topologically by a closed Möbius band, the latter can be described as a punctured projective plane (Seifert and Threlfall, 1980, p.11).

7 A connected sum is the joining up of distinct manifolds at the place where each has a disk cut out, so that the remaining boundaries can be 'zipped up' together.

8 Different cuts produce different effects in the spaces we have examined so far. Thus, cutting a Möbius band alongside its middle (a median cut) transforms this one-sided non-orientable surface into a two-sided orientable one, while applying the same but close to one of the edges splits the Möbius band into another Möbius band (non-orientable), interlinked with an orientable two-sided loop (Hughes, 2013, p.82). In Lacan's words, the property of the Möbius band is 'nowhere other than precisely in the cut, which is the only thing that has the shape of the Möbius strip' (Seminar XII, cited in Hughes, 2013).

9 It is interesting to note that, when it comes to higher dimensions, many mathematical problems that can be solved at dimensions of five and above have no solution in four dimensions (Mikhailov, 2004).

References

Adams, C.C. (2000). *The knot book: An elementary introduction to the mathematical theory of knots*. Providence, RI: American Mathematical Society.
Alexandroff, P. (1961). *Elementary concepts of topology*. New York: Dover Publications.
Armstrong, M.A. (1983). *Basic topology*. New York: Springer-Verlag.
Arnold, B.H. (2011). *Inutititve concepts in elementary topology*. New York: Dover Publications.
Banchoff, T. (1996). *La quatrième dimension: Voyage dans les dimensions supérieures*. Paris: Pour La Science Diffusion Belin.
Barr, S. (1964). *Experiments in topology*. New York: Dover Publications.
Burgoyne, B. (2018). The changing forms of a research programme. In Bailly, L, Lichtenstein, D. and Bailly, S. (eds.). *The Lacan tradition: Lines of development – Evolution of theory and practice over the decades*. London: Routledge, pp.46–89.
Bursztein, J.-G. (2017). *Subjective topology: A lexicon*. Paris: Hermann.
Casati, R. and Varzi, A.C. (1994). *Holes and other superficialities*. Cambridge, MA: MIT Press.
Charraud, N. (2004). La topologie «TBMCC» de Jaques Lacan. In Cartier, P. and Charraud, N. (eds.), *Le réel en mathématiques: Psychanalyse et mathématiques*. Paris: Agalma, pp.135–151.
Christensen, W.J.Jr. (2017). God is a porcupine – Brain, consciousness and spacetime physics. *Journal of Modern Physics*, 8, pp.1294–1318.
Coetzee, J.M. (1983) *The life and times of Michael K*. London. Secker & Wearburg.
Eigen, M. (2012). *Kabbalah and psychoanalysis*. London: Karnac.
Epple, M. (1999). Geometric aspects in the development of knot theory. In James, I.M. (ed.). *History of topology*. Oxford: Elsevier, pp.301–357.
Freud, S. (1898). Sexuality in the aetiology of the neuroses, SE3. pp.261-285.

Freud, S. (1915). The unconscious. SE14, pp.161–215.
Greenshields, W. (2017). *Writing the structures of the subject: Lacan and topology*. London: Palgrave Macmillan.
Hilton, P. (2008). A brief, subjective history of homology and homotopy in this century. *Mathematics Magazine*, 61(5), pp.282–291.
Hughes, T. (2013). The Klein bottle. *The Letter*, 53, pp.57–85.
Kasner, E. and Newman, J. (1940). *Mathematics and the imagination*. New York: Dover Publications.
Lacan, J. (1965). *Les problèmes cruciaux pour la psychanalyse*. Publication hors commerce.
Lacan, J. (1972). L'Étourdit. In Lacan, J. (ed.). (2001). *Autres ecrits*. Paris: Éditions du Seuil, pp.449–495.
Lacan, J. (1974). Seminar XXI, *Les non-dupes errant*. Unpublished.
Lacan, J. (1992). *The ethics of psychoanalysis. Book VII*, 1959–1960. London: Routledge.
Lacan, J. (1997 [1981]) *The seminar. Book III. The psychoses, 1955–1956*. London: WW Norton.
Lacan, J. (2020). *L'identification: Séminaire 1961–1962*. Éditions de l'Association Lacanienne Internationale. Paris: Publication hors commerce.
Laurent, É. (2019) – Interpretation: From truth to event.
Magee, B. (2016) *Ultimate questions*. Oxford: Princeton University Press.
Mendelson, B. (1962). *Introduction to topology*, 3rd ed. New York: Dover Publications.
Mikhailov, R.V. (2004). On some questions of four dimensional topology: A survey of modern research. *Hypercomplex Numbers in Geometry and Physics*, 1, pp.100–103.
Miller, J.-A. (1996) – Interpretation in reverse.
Nasio, J.-D. (2010). *Introduction à la topologie de Lacan*. Paris: Petit Bibliothèque Payot.
Poincaré, H. (1912). Pourqui l'espace a trois dimensions. *Revue de métaphysique et de morale*, 20(4), pp.483–504.
Seifert, H. and Threfall, W. (1980). *A textbook in topology*. London: Academic Press.
Sogyal, R. (2002). *The Tibetan book of living and dying*. London: Rider.
Stewart, I. (2002). *Flatterland: Like Flatland, only more so*. New York: Basic Books.
The Open University (2016). *Surfaces*, Open Learn. www.open.edu/openlearn.
Tucholsky, K. (1931). The social psychology of holes.
Vandermersch, B. (2008). *Le cross cap du Lacan ou "asphère"*. Paris: Association lacanienne internationale.
Vanheule, S. (2020). On a question prior to any possible treatment of psychosis. In Hook, D., Neill, C. and Vanheule, S. (eds.). *Reading Lacan's Écrits: From 'The Freudian thing' to 'Remarks on Daniel Lagache'*. London: Routledge, pp.163–205.
Vivier, L. (2004). *La topologie: L'infini matrisé*. Paris: Le Pommier.
Volkert, K. (2013). Projective plane; A history. *Bulletin of Manifold Atlas*, pp.1–4.
Weeks, J.R. (2002). *The shape of space*. 2nd ed. London: CRC Press.
Wells, D. (1991). *The Penguin dictionary of curious and interesting geometry*. London: Penguin Books.
Yalom, I.D. (1980). *Existential psychotherapy*. New York: Basic Books.
Zeeman, E.C. (1966). *An introduction to topology. The classification theorem of surfaces*. Mathematics Institute, University of Warwick.

Chapter 8

The unconscious as knots

The relevance of knots to this exploration of the unconscious as space is two-fold. First, in terms of progression of the mathematical thinking in topology, knot theory is one of the most recent and revealing developments. Second, psychoanalytically, knot theory is where Lacan more or less left his work at the end of his life, with an implicit invitation to the next generation to take this further.

Greenshields (2017) emphasises that, for Lacan, the function of the knot is not to amount to 'a grand synthetisation and completion of psychoanalytic theory' (p.9), but rather to be 'deployed as the nonsignifying support of that which cannot be theorised' (*ibid.*). In this respect, knots hold a particular relationship to the unknown and impossibility. When the unconscious is regarded as the impossible, what knots offer is a writing that presents not the content, but the structure of 'the impossible to say' (Greenshields, 2017, p.23). Like Lacan himself, Greenshields insists time and again that topology is not a metaphor, but the very structure at stake: 'The unconscious *is* a topology and its topology is that of a knot' (Greenshields, 2017, p.25), with the topological structures at play seen as belonging to the order of a *hypokeimenon*, constituting not a mere representation, but the actual support, the very substance of the psychoanalytic field itself (SXIII, 1966, p.160).

In his exploration of four-dimensional space as a cultural object, Blacklock (2018) identifies the knot as 'an object that is a thing and idea, formal and material, mediator and terminus' (p.9). In the early days of mathematical thinking around the notion of four-dimensional space and knots, much of the debate, as he shows, became derailed by arguments around spiritualism. It is important not to forget that, in the incipient days of knot theory, Peter Tait's classification of knots was against the background of an understanding of the world defined by Lord Kelvin's notion that atoms were knots in ether (Epple, 1999).

If we take seriously the assertion that topology is the space of analysis itself, it follows mathematically that this type of intervention is intrinsically equipped to engage with the unconscious understood as topological object. In between the two is the space of language, equally topologically structured, an intermediate space in which movement takes place along knot-like pathways.

8.1 Knots – a primer

Like topology, as rooted in Poincaré's *Analysis Situs* of 1895, knot theory also developed in the pursuit of answers that related to questions from other branches of mathematics, as a new way to address what could not be answered with what was already known (Epple, 1999; Sarkaria, 1999). At the same time, as an understanding of knots and their complements (i.e. of the rest of the space, where the knot is not) developed in the first part of the 20th century, the links to topology and manifolds became increasingly strong, to the extent that knot theory never quite became a 'self-sustaining subfield of mathematics' (Epple, 1999, p.331), but remained a subfield of topology.

Historically, knot theory emerged as a precursor to Mendeleev's table of elements, as an early model of atomic structure, at a time when it was believed that ether pervaded all space and matter was understood as knots in it (Adams, 2000, p.5). One hundred years later, the relevance of knots returned, in relation to understanding the structure of DNA.

This domain studies knots and links – a concept never referred to as such in Lacanian analysis, but much used, given that what Lacan called a Borromean knot is actually, mathematically, a link.

Given that much of the work involving knots was a task of classification and inscription, and given that this relied largely on diagrams, it is easy to forget that knots are 'genuinely three-dimensional objects' (Epple, 1999, p.302), and not just a kind of writing. Also, the knots of mathematics are different from those encountered in everyday life in that they have no loose ends, but rather the two are spliced into a continuum. In this sense, a knot is 'a closed curve in space that does not intersect itself anywhere' (Adams, 2000, p.2).

More specifically, a knot is a subspace of Euclidean three-dimensional space (Armstrong, 1983, p.213), consisting in a cycle, or a closed loop, homeomorphic to a circle (Vivier, 2004, p.147). The circle is, by definition, the trivial knot, known as the unknot, and has some kind of equivalence to the place zero has among numbers. What makes it trivial is that it lies in a plane (Crowell and Fox, 1963, p.5), whereas all other knots require an extra dimension.

In mathematics, knots are represented as projections, two-dimensional drawings with a certain number of crossings (i.e. places where the cycle crosses itself, where one section of the knot passes under another section of itself). The same knot can have a multitude of projections, which can be deformed into each other without cutting. The projection of interest is the one with the smallest possible number of crossings. As succinctly put by Adams, '[M]uch of knot theory is concerned with telling which knots are the same and which are different' (Adams, 2000, p.4), and indeed with discerning whether a projection is that of a knot, or just a jumbled-up circle (unknot). In other words, the primary task is to distinguish non-trivial structures and to identify equivalent ones.

The notion of homeomorphism employed in knot theory appears to be, at first sight, different from the one employed in the classification of topological spaces. So far we have employed homeomorphism to refer to the transformation of one surface into another without cutting, merely by stretching, bending, or shrinking it. In the case of knots, another kind of homeomorphism seems to be at play, in that cutting is permitted before the path is stretched, bent, straightened, and then 'glued' back again as it was just before the cut. It is in this second sense that the trefoil knot is homeomorphic with the circle. The difference between the two notions resides in the fact that for knots the homeomorphism refers strictly to the space of the knot and not to the embedding space. In other words, every point of the trefoil knot can be mapped continuously to and from a circle, but the same is not true for the space in which these two sit (Seifert and Threlfall, 1980, p.2). Also, in three-dimensional space we can construct a left-hand and a right-hand trefoil knot, and these cannot be turned into each other without cutting. Most importantly, '[T]he property of being knotted is not an intrinsic topological property of the space consisting of the points of the knot, but is rather a characteristic of the way in which that space is embedded in R^3 [three-dimensional space]' (Crowell and Fox, 1963, p.4). Thus, knottedness is derived from embedding in three-dimensional space, so that, mathematically, a knot is an embedding of a circle that cannot be deformed continuously into an unknot (Stewart, 2013, p.100).

We arrive in this way back to the question of four-dimensional space. A knot and a circle differ only in the manner in which they are embedded in three-dimensional space. If, however, we consider this to be merely a subspace of four-dimensional space, then any knot and the circle can be continuously deformed into one another without cutting and without self-intersection (Seifert and Threlfall, 1980, p.3). In other words, there are no knots in four-dimensional space; yet again, we find that impossibility is a consequence of limited spatiality in terms of dimensions. What is impossible to untie in three dimensions can be untied in the fourth. A trefoil knot, for instance, can be freed into becoming a circle by moving one of the overlapping segments along the fourth dimension so that the two segments are no longer in the same place at the same time. What is above, to start with, can be behind. Furthermore, a curve cannot be knotted in four dimensions, but a surface can (Wells, 1991, p.133). In the words of Felix Klein, the mathematician of the 19th century, of the Klein Bottle fame, 'a knot is a property of a closed curve in three-dimensional space, but in four-dimensional space the closed curve can be unknotted by deformation' (Klein, 1979, p.157).

A knot can be thought about in relation to a surface in two different ways: as a writing on a particular support or as a three-dimensional mathematical object which is 'surrounded' by a surface that is moulded to its contours. Thus, when we think of a knot k, we implicitly consider the space it sits in. In three-dimensional space, the complement of the knot is the entirety of this space less the space occupied by the knot itself, R^3-k. A knot can 'sit' in a sphere or it can be contained in a torus (which appears itself knotted as it traces the shape of the knot, as it surrounds its pathway – Adams, 2000, p.85). One way to visualise this is by thinking of the mould filling technique in sculpture, with the knot being surrounded by the solidity of the cast as the space against which it is defined and in which it is held.

Mathematically, a knot designates 'a path in space that begins and ends at the same point' (Totaro, 2008, p.385). Some knots can be turned into others, others cannot. Some knots are made up of a finite number of line segments (tame knots); others are constituted as an infinity of knots in sequence (wild knots). Most of the existing knot theory is concerned with tame knots. Since mathematically knots are dealt with in their projected form, or what Lacan called a *mise-à-plat*, their representation can amount to a kind of writing. Mathematicians call this a 'nice projection', one which has a finite and minimal number of crossings (no more than two points of the knot are mapped in the image of the knot – Armstrong, 1983, p.215). For any knot k, all its possible 'shufflings' that are homeomorphic to it constitute the knot group of k. It is helpful to think of knots as bits of looped rope which, when twisted from one position to another, are deformed in a way that creates a one-to-one correspondence between the points in each of the two positions.

The continuous deformations possible that can be used to simplify knots until no further simplification is possible were formalised in 1927 by a German mathematician, Kurt Reidemeister, and are known as the Reidemeister moves (Figure 8.1). The 'disentangling' of a knot aims to bring it to what is known as a regular position, that is to say, to a state where the only multiple points are double points and each of these depicts a genuine crossing (Crowell and Fox, 1963, p.6). Each of these double points of the projected image of a knot (the *mise-à-plat*) into a crossing point is the image of two distinct points of the knot. The one above is the over-crossing and the one below is the corresponding under-crossing. It is from this representation that knots can be classified, as they are compared and distinguished or found identical, on the basis of so-called knot invariants, that is to say, measures associated with a knot that does not change under continuous deformations of that knot.

The rearranging of the 'string' of the knot happens in three-dimensional space, without allowing it to pass through itself. Notice that, in order for the 'writing' of the knot to be transformed, access to a dimension outside the space where the inscription sits is necessary. In other words, in order to shuffle it on the page one needs to lift the knot in the space above the page on which it is inscribed, before setting it down again in a new flat form of inscription. It is also possible to deform the knot by stretching the surface on which it is written in a way that we recognise from the rubber-sheet geometry of topology.[1]

Reidemeister moves offer three ways in which a writing (projection on a plane) of a knot can be modified in such a way that the relation between the crossings is modified towards simplification: adding or taking out a twist, adding or removing two crossings, sliding one strand from one side to the other side of a crossing. Each transformation of this kind changes only the projection of the knot, making it easy to read, but it does not modify the knot it represents (Adams, 2000, p.14).

Stewart (2013) explains these operations in simple terms: 'Each move can be carried out in either direction: add or remove a twist, overlap two strands or pull them apart, move a strand through the place where two others cross' (p.101). Given

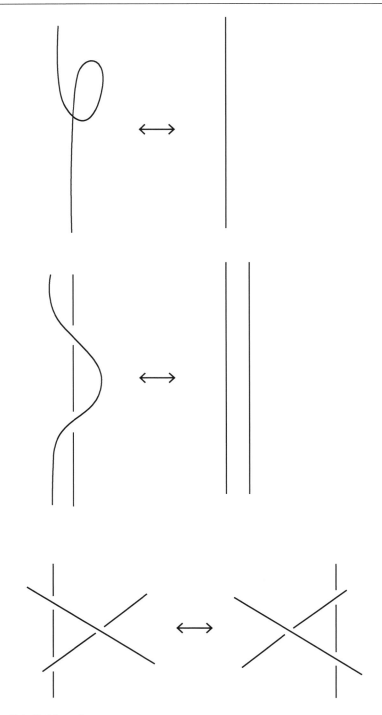

Figure 8.1 Reidemeister moves.

that knots are viewed as equivalence classes, with each knot supporting a multitude of diagrammatic representations, diagrams that could be deformed into each other by a finite sequence of such moves represent the same knot (Epple, 1999, p.307). The main preoccupation of early knot theory was the establishment of invariants, namely, of parameters that can uniquely identify one knot as fundamentally distinct from another, as table knots were created.

Interestingly, this early formalising in knot theory was more or less contemporaneous with Freud's publication of *Inhibitions, Symptoms and Anxiety* and *Civilisation and Its Discontents*, at a time when psychoanalytic technique was not his primary preoccupation. Nevertheless, the diagrams of these moves are reminiscent of his earlier work about the way in which the speech of the patient approaches and moves away from the pathological nucleus, describing a pathway in the unconscious, traced in language (Freud, 1895, p.288). If, as Lacan did later, one thinks in terms of chains or strings of signifiers, these moves used to simplify knots are also direct representations of key psychoanalytic interventions. What these moves do to knots is to change the way in which a given entity is embedded in its space, so that its specific identity can be recognised and potentially modified. The untangling from false loops or removal of layers of empty speech, the creation of new possibilities and the re-positionings at play are fundamentally moves that characterise the analytic experience at every scale: within the session, between sessions or over the entirety of an analysis. Note that these are interventions that do not require a cut, which explains why some degree of change is also possible in other clinical interventions, which do not engage with the more complex technical Lacanian take on analytic work.

8.2 Knots as structure and pathways through language

Lacan regards knot theory as a way of grasping something about the difficulty with representing even the most simple of combinations (Lacan, 2020, p.230). According to Charraud, with the move from the topology of surfaces to the use of knots, Lacan marked a shift towards a new question, away from the emergence of meaning towards the emergence, out of a knot of meaning, of the object towards which desire tends without ever obtaining it (Charraud, 1986, p.7). This is the object-cause-of-desire, placed by the subject in the torus hole, in an attempt to both plug the gap of the lack and keep the desire, as engine of life, going. In her view, this marks a particular representation of the notion of (topological) convergence towards the absence of the signifier[2] (Charraud, 1986, p.8). In other words, understanding something about knots allows us to grasp the structure of the territory outside the realm of meaning, a territory crucial for what is at stake in our search for satisfaction on the side of life.

The interplay between the three registers of experience – the Real, the Symbolic and the Imaginary – is Lacan's most prominent development rooted in knot theory. One of the problems with the common representation of the Borromean knot is the sense of symmetry it conveys among the three registers. Yet, more subtle distinctions are both possible and necessary. For instance, Bursztein (2017) refers to this structure as the fabric or material of subjectivity, 'insofar as it realises itself in a substance within which the subject of the unconscious unfolds itself' (p.17). These interconnected registers can be thought of as dimensions of the space of subjective experience, with each dimension being linked to one particular set of numbers, or a particular signifying density. Thus, Bursztein links the Real to the set of real numbers, the symbolic to the set of irrational numbers and the imaginary to the set of integers, and proposes a fourth dimension pertaining to the phantasy, which he places in correspondence with the domain of complex numbers (Bursztein, 2017, p.47). This differentiation offers both a sense of coherence and of radical difference among the components of this space. A more detailed exploration of the differences at play is offered by Bristow (2021), who considers the continuous nature of the Real, as opposed to the discrete, fragmented, sequence-like nature of the Imaginary and the Symbolic. By moving away from the image, it is easy to recognise that no one-to-one mapping between the three registers is possible. A similar distinction is made by Vandermersch, who places the signifier in the Symbolic and the letter in the Real. The move from the conscious to the unconscious amounts to a change of topology (Vandermersch, 2009, p.141). This needs to be understood as a move between radically different spaces, between which no direct or complete mapping can exist. So, each register is both a space in its own right and a dimension at the level of subjective experience.

Recall the various pathways possible on the torus, including what Lacan called the interior eight, a non-trivial trajectory which announces the notion of a knot, without being one (Vappereau, 2006). Taking the state of the letter in the unconscious as represented by this inverted eight – identical to itself and different from itself – and distinguishing between surfaces on which such a loop can be inscribed and those which do not support it, the transition between topological spaces occurs at the level of phantasy. As we have seen, the topological space of the phantasy is the cross-cap which can be cut into a Möbius band and a disk, each of them bordered by the same inverted eight, which is homeomorphic to a circle. The immersion of such a surface in three-dimensional space entails a line of self-intersection, a line akin to a hole which creates a kind of boundary or cut between spaces of incompatible valences. This boundary of the inverted eight is bivalent – both double and simple, pertaining both to the Möbius band and to the disk. It is here that Lacan situates the phallus, as 'a signifier that the speaking subject searches for in pursuit of that which causes desire' (Vanheule, 2020, p.186).

8.3 Knots in the clinic

It can be quite difficult to see the immediate link between such abstract mathematical concepts and the day-to-day work in the consulting room. It becomes much easier, however, if we keep in mind three important notions: first, the space of experience is defined structurally by impossibilities (holes); second, speech and more generally chains of signifiers constitute pathways or trajectories of movement across the space of experience; third, the work of analysis addresses the former via the latter; that is, by modifying the pathways, it offers new and more satisfying ways of navigating and engaging with impossibility.

And now for a closer look at the specific relevance of knots within the reality of topology. We can distinguish here between knots proper and the so-called Borromean knot, which is most prominent in the Lacanian literature. As far as proper knots are concerned, the most important aspect is that some of them constitute pathways on surfaces with holes, e.g. one of the simplest knots, the trefoil knot, is a pathway on a torus. Intervening analytically in a way that opens up a stuck chain of signifiers into a different kind of pathway can offer a radical change in the way the subject navigates the holed space of their existence.

Helpfully, Vandermersch (2009) clarifies that the chains of signifiers operate like Markov chains (p.137). In mathematics, these are systems where transitions from one state to another cannot happen in any odd way, but are constrained by the current state and by time. Any next step is constrained by the point from which the step is taken. This makes it very easy to see how analysis works by cutting the chain, so that the range of possible steps, and therefore of future states, becomes modified. Sometimes a cut is needed, sometimes interventions akin to the Reidemeister moves suffice, and most likely both will occur several times over the course of any analysis (see Figure 8.1).

In essence, the moves change the relation between the crossings. Adding or removing a twist creates a space that can circumscribe something, constituting an operation at the level of enjoyment. Note that access to a third dimension is needed for the creation of a loop which is not merely a trivial bending of the pathway. One needs to come off the flat space of the page in order to produce a significant effect. In the case of adding or removing crossings, we are dealing with separating distinct elements that seem like one or bringing together things that seem separate. Finally, in the case of sliding, the effect is that of changing the relationship between signifiers in terms of before/after, introducing a change in the structure of a narrative. Mathematically, such moves transform a knot without cutting, making it possible to simplify it and therefore classify it. Clinically, this can be understood in terms of the effects of interpretation, which is not about how individual analysts work, but about how analysis itself operates. Interpretation – even when amounting to a cut – does not add to or eliminate from the signifiers brought by a patient, but rather facilitates and generates a reconfiguration of their relative positions, revealing what is already there, twisted, hidden, obscured by something else, and changing the relative positions of signifiers, opening up new routes and perspectives. Furthermore, interpretation that cuts changes the topology in such a way as to make visible the space of desire for the subject and the nature of this space (Cochet, 1998, p.104).

Thus, knots tell us something about how and why analysis works. If we apply the rigour of knot theory to chains of signifiers, on the basis of equivalence classes, analytic interventions are homotopic, that is to say, they are paths that can be deformed into each other. Therefore, although each analyst intervenes in their own way, they all do it on some path between the position with which a patient arrives and the space where the subject of the unconscious emerges. Such understanding is consistent with a movement away from the production of meaning, towards a recognition of how analysis operates through the effects of interpretation, with the analyst as a mere operator or intermediary of analysis itself. The analyst does all of this by opening up access to an additional dimension.

From a three-dimensional structure, the Borromean knot has been transformed in Lacanian analysis into a kind of flat map of the interplay between the registers of experience and manifestations of desire. This structure demarcates a space within which 'the topological space of the subjective structure realises itself' (Bursztein, 2017, p.49). The nature of this subjective space is not the same with that of the physical space inhabited by the body; it, rather, is what Lacan called *n'espace* (Bursztein, 2017, p.51).

Before we consider this closer, it is important to distinguish conceptually between borders and boundaries. In topology, the existence of a boundary is a structural feature of a space, defining of it. The boundary number is one of the invariants involved in the classification of manifolds. In contrast with this, a border is a constructed partitioning of a space, a kind of fence built within the space, such as a Jordan curve, which delimits the inside from the outside, splitting the space in two. A cut can have the quality of a localised separating partition, but it also can alter the structure of the entire space when it creates a new boundary.

The partitioning of the space delineated by this Borromean structure works only in a *mise-à-plat* scenario, where the links are not seen as more that an inscription on a flat surface, unlike the actual mathematical object, where the 'holes' are in a relation of continuity with each other, rather than just three conjoined Jordan curves.[3] Equally, the projected inscription in the plane creates some false points of apparent intersections between registers, which are actually separate points in three dimensions. Bursztein gets around this by mentioning the pretzel-like topology of the space in question, regarding the Borromean structure as an n-holed torus. In this kind of inscription, Bursztein (2017) offers a construction centred on the space of object-cause-of-desire, obj a, surrounded by three types of jouissance, which in turn are surrounded by three aspects of the subjective body, one for each register (pp.63–64). In this configuration, the boundary of the area of *obj a* operates like a Möbius band which lends a dynamic character to the entire structure, with the psyche constantly moving between lack and phantasy, between chronological time and moments outside it, between lived time and the timelessness of the unconscious. This makes the unconscious into a space of potentialities whose nature escapes complete representation. As we have seen in Chapter 5, representation, spatially, is a question of dimensionality.

8.4 Knots and dimensions

Our exploration of topology so far has shown us that certain topological properties are intrinsic and therefore preserved under all continuous transformations, whereas others depend upon the placement in space of a set of points in three-dimensional space, and are preserved only under the mapping of the whole space. We are back to the question of embeddedness, once again.

In the way that a circle can be knotted in three-dimensional space, a sphere can be knotted in four-dimensional space (Weeks, 2002, p.193). However, knottedness is not a property of the knot by itself, but of the way in which it is embedded. That is why a knot can be undone in four-dimensional space as it is the space dimensionality that is different, not the essence of the knot itself. Zeeman urges his reader to think of surfaces as abstract objects that exist on their own, without being embedded in anything (Zeeman, 1966, p.8).

In an afterword updating an ongoing debate over the nature of space and time, two scientists at the cutting edge of research on the matter, Stephen Hawking and Roger Penrose, agree that the emerging understanding of space is 'a picture in which the overall spatial geometry is very close to flat' (Hawking and Penrose, 1996, p.140). At the same time, they also recognise the importance of the notion of 'brane worlds', 'in which what we experience as "physical reality" may actually be some kind of boundary of a higher-dimensional structure' (Hawking and Penrose, 1996, p.141).

The space that knots embody has a particular nature, as they can operate as boundaries between what can and cannot be articulated in language or represented in images. Mathematically, some spaces are bounded by knots, which constitute the demarcation of an impossibility or the possibility of uninterrupted movement, depending on the dimensionality of the space in which they are encountered.

This resonates with the analytic reading of subjective experience as occupying such a space, a kind of littoral (Vandermersch, 2009, p.137), a coastline around the edges of experience, whose essence can be captured either poetically or mathematically. Lacan moved away from the former to the latter approach, over the unfolding of his theorising, so that, as he came to formulate more of the Real unconscious, in the latter part of his work, he emphasised: 'The real of which I speak is absolutely unapproachable, except by way of mathematics' (Lacan, 1972, p.49, *my translation*). As Cléro (2010) explains, the Real is to be found in mathematics rather than in an empirical way (p.14).

In Lacanian analysis, the lack has a structuring function. In this sense, the unconscious is neither at the level of being nor at that of non-being, but of the unrealised (Lacan, 1979, p.30), and as such remains outside contradiction, spatio-temporal location and the function of time (Lacan, 1979, p.31).

In his last seminar, Lacan declares himself as having been wrong to use the Borromean knot as a way of capturing something about the interplay between three radically different registers of experience, as he acknowledged that there is nothing that supports the Imaginary, the Symbolic and the Real (Hughes, 2013, p.66). It is unclear to what extent this reconsideration is a spatial one.

Physics offers its own specific views of space and the psyche, of which one in particular deserves to be mentioned. I refer to the hypothesis explored by Christensen (2017) that spacetime is a field of a gravitational type, and that the brain – although in itself not physically complex enough to perform all the functions and operations it is capable of – has evolved as a composite of facilitators capable to perceive quanta from this field, and able to 'interact with microscopic spacetime to produce and maintain our rapidly shifting mind, which interprets through the brain, the external world, the body it resides in, and awareness of its conscious self' (p.1312).

This notion of space is significantly removed from the view of space as the container of the world and of time as a fourth dimension. Since Einstein's relativity theories modified our understanding of the world we live in way beyond Newton's conception, physicists seem to agree that '[S]pace does not exist independently from time' (Rovelli, 2017, p.59). This modification does away with the distinction between space and time and particles that make up the world, replacing it with the notion that the world is simply a combination of fields and particles. In this sense, space becomes 'one of the "material" components of the world' (Rovelli, 2017, p.65). This is a space which curves and ripples, and its posited attributes combine the notions of four-dimensional space with those of topology.

This particular reading of quantum theory also does away with the notion of a continuum. Space is a field that manifests in quanta, it is not infinitely divisible into smaller and smaller points. Indeed, this is a space defined by granularity, indeterminism and relationality. Granularity refers to the limits of divisibility of space to very small but finite units – quanta. Indeterminism refers to the absence of determination in the state of quanta in between moments of interaction. Relationality is the quality of a universe where things do not exist in an absolute way, but in relationship to each other: electrons materialise where they collide with something else, not unlike the subject, who operates not in a continuum, but in movement from one signifier to another, and is defined in relation to a particular kind of other, as Lacan carefully showed. Seeing the electron as 'a combination of leaps from one interaction to another' (Rovelli, 2017, p.101) is entirely congruent with the notion of a Lacanian subject. In this conception of space, no variable of an object is defined as fixed between an interaction and the next, but it remains undetermined in a spectrum of possible values, each with their attached probabilities. 'It is only in interactions that nature draws the world' (Rovelli, 2017, p.115). In other words, it is not objects but events that constitute the world we live in.

Interestingly, quantum mechanics is considered to have been born with Max Plank's work in 1900, the same year that saw the publication of Freud's *Interpretation of Dreams*. The most recent version of this domain of theoretical development, known by the name of loop quantum gravity, is posited as a main alternative to the relatively older string theory. The loop of the name relates to the closed lines that were found in the 1980s as solutions to the equations of space as gravitational field. Moreover, the quanta of this field are not in space, but are themselves space: 'Space is created by the interaction of individual quanta of gravity' (Rovelli, 2017, p.148). Their interaction also leads to the emergence of time, as a localised phenomenon (p.153). Such fields that do not need a spacetime substraturm, but generate their own spacetime, are called covariant quantum fields. But what perhaps is more striking, most at odds with the Freudian view of equilibrium as pursued in the immovability of death, is that the current picture of the world offered by physics is not one of stasis, but of perpetual movement: 'The impossibility of anything being entirely and continuously still in a place is at the heart of quantum mechanics'

(Rovelli, 2017, p.198). This is a world without time, which becomes merely 'an effect of our overlooking of the physical microstates of things. Time is information we don't have. Time is our ignorance' (Rovelli, 2017, p.223).

All these notions put into question everything we have considered thus far about space, dimensions and change, taking us all the way back to the old question of how a line can have one dimension, when it is made out of infinitely large number of points with no dimensions, posed by Democritus before the Euclidean view became the norm. Indeed, one is put in the mind of geoglyphs as the kind of writing one can be inside, occupying the same plane with it, and walking around it as if in a maze, without seeing what it inscribes. The need for a higher dimension from which to apprehend the inscription is intrinsic to the pursuit of change and analysis is the closest to a solution to the conundrum of being inside and out of language at the same time.

Some clinical considerations, in the remaining two chapters, will make it possible to bring together all these strands, and to open the question for work yet to be undertaken in developing further the ideas sketched here.

Notes

1 The former transformation is called ambient isotopy and the latter planar isotopy.
2 This clarification might also elucidate why Lacan chose to refer to the Borromean structure as a knot. Topologically, it is not a knot, but a link, so closer to a chain, a term he had already cemented with reference to signifiers and which no longer could, perhaps, serve to mark this radical departure from language to topology.
3 Recall, from Chapter 6, that a Jordan curve is a continuous curve which divides the plane into two regions, an interior and an exterior, where the interior cannot be reduced to nothing, as it contains a circle of finite radius (Crilly, 1999, p.10).

References

Adams, C.C. (2000). *The knot book: An elementary introduction to the mathematical theory of knots*. Providence, RI: American Mathematical Society.
Armstrong, M.A. (1983). *Basic topology*. New York: Springer-Verlag.
Blacklock, M. (2018). *The emergence of the fourth dimension: Higher spatial thinking in the fin de siècle*. Oxford: Oxford University Press.
Bristow, D. (2021). *Schizostructuralism: Divisions in Structure, Surface, Temporality, Class,* Abingdon: Routlegde.
Bursztein, J.-G. (2017). *Subjective topology: A lexicon*. Paris: Hermann.
Charraud, N. (1986). Problematique autour de la topologie. *La Lettre Mensuelle de l'École de la Cause Freudienne*, no.46.
Christensen, W.J.Jr. (2017). God is a porcupine – Brain, consciousness and spacetime physics. *Journal of Modern Physics*, 8, pp.1294–1318.
Cléro, J-P. (2010). L'utilité des mathématiques en psychanalyse: Un problème de chrestomathie psychanalytique. *Essaim*, 24, pp.7–36.
Cochet, A. (1998). *Lacan geometre*. Paris: Anthropos.
Crilly, T. (1999). The emergence of topological dimension theory. In James, I.M. (ed.). *History of topology*. Oxford: Elsevier, pp.1–24.

Crowell, R.H. and Fox, R.R. (1963). *Introduction to knot theory*. New York: Dover Publications.
Epple, M. (1999). Geometric aspects in the development of knot theory. In James, I.M. (ed.). *History of topology*. Oxford: Elsevier, pp.301–357.
Freud, S. (1895). Studies on hysteria. SE2.
Greenshields, W. (2017). *Writing the structures of the subject: Lacan and topology*. London: Palgrave Macmillan.
Hawking, S. and Penrose, R. (1996). *The nature of space and time*. Oxford: Princeton University Press.
Hughes, T. (2013). The Klein bottle. *The Letter*, 53, pp.57–85.
Klein, F. (1979). *Development of mathematics in the 19th century*, trans. M.Ackerman. Brookline: Math Sci Press.
Lacan, J. (1972). *Seminar XIX, Ou pire: Le savoir du psychanaliste, 1971–1972*. Unpublished.
Lacan, J. (1979[1973]). *The four fundamental concepts of psychoanalysis*. London: Penguin Books.
Lacan, J. (2020). *L'identification: Séminaire 1961–1962*. Éditions de l'Association Lacanienne Internationale. Paris: Publication hors commerce.
Rovelli, C. (2017). *Reality is not what it seems: The journey to quantum gravity*. London: Penguin Books.
Sarkaria, K.S. (1999). The topological work of Henri Poincaré. In James, I.M. (ed.). *History of topology*. Oxford: Elsevier, pp.123–167.
Seifert, H. and W. Threfall (1980). *A textbook in topology*. New York: Academic Press.
Stewart, I. (2013). *Seventeen equations that changed the world*. London: Profile Books.
Totaro, B. (2008). Algebraic topology. In Gowers, T. (ed.). *The Princeton companion to mathematics*. Oxford: Princeton University Press, pp.383–396.
Vandermersch, B. (2009). Littoral ou topologie du refoulement. *La revue lacanienne*, 1(3), pp.137–144.
Vanheule, S. (2020). On a question prior to any possible treatment of psychosis. In Hook, D., Neill, C. and Vanheule, S. (eds.). *Reading Lacan's Écrits: From 'The Freudian thing' to 'Remarks on Daniel Lagache'*. London: Routledge, pp.163–205.
Vappereau, J.M. (2006). La D.I. [http://jeanmichel.vappereau.free.fr/textes/La%20DI.pdf].
Vivier, L. (2004). *La topologie: L'infini matrisé*. Paris: Le Pommier.
Weeks, J.R. (2002). *The shape of space*. 2nd ed. London: CRC Press.
Wells, D. (1991). *The Penguin dictionary of curious and interesting geometry*. London: Penguin Books.
Zeeman, E.C. (1966). *An introduction to topology. The classification theorem of surfaces*. Mathematics Institute, University of Warwick.

Part IV

Clinical implications

What is central to the fundamental understanding of being human that comes from psychoanalysis, is the recognition that access to ourselves by way of our own unconscious is always through another. This someone other than oneself operates as a proxy for the missing dimension which makes it impossible to have direct and unmediated access to something of our own being that is experienced as both internal and out of reach. We look for something of ourselves in and through others. It is no surprise then that everyone who arrives to analysis brings some suffering in the field of relationships and attempts to resolve this in the space of yet another relationship, by addressing a special kind of Other, the analyst as the perceived embodiment of the Lacanian subject-supposed-to-know (Lacan, 2020, p.19). In this sense, crucially, it is not the person but the function of the analyst that is in operation in the cure. This proxy solution of addressing the Other in the analyst is always destined to remain imperfect, as this mediated access to oneself is always fragmented and lengthy, with time taking the function of the missing spatial dimension, to some extent. In trying to resolve their experience of impossibility, the analysands find, instead, a freer way of being with it. Structurally, the problem of degrees of freedom impacts upon both the intervention on the side of the analyst, and the possibility of change, on the side of the analysand: we try to intervene from the inside as if from the outside, except neither exists as such, as we are always immersed in the experience, analyst and analysand alike. Whatever is not entirely ruled out by impossibility, can only be dealt with in a fragmented, disjointed way, from a lower towards a higher dimension where the joining up happens, in the fundamentally out-of-reach continuum of the unconscious.

Although the analysand addresses the analyst as the embodiment of the knowledge they themselves both seek and hold, the one thing that the analyst definitely knows is that they do not hold that particular knowledge themselves. At the same time, this is not to say that the analyst has no knowledge (Evans, 1996). Indeed, if the analyst knows anything, it is that their function is to allow the analysand to find their own knowledge and way to their subjectivity. The central question at this point on our journey is how all the mathematical elements explored so far can inform and illuminate the clinic, how do they sit in the landscape of knowledge of the analyst. It is, hopefully, easy to see how some elements of mathematics and topology are crucial to their capacity to hold well their own knowledge and impossibility in relation to the unconscious as space.

It should be clear by now how a familiarity with notions of spatial dimensions and the topology of holed spaces makes it possible to grasp something about what is at stake in any analytic encounter: the nature of suffering, as well as why and how analysis works to address this. Using this particular understanding of structure, we can give a more accurate expression to the impossibility that underpins all suffering and also grasp how analysis itself operates in spatial terms.

What we have considered so far is a rigorous way of thinking about the impossibility at the core of subjective discord in relation to the fragmented and often incomprehensible experience of our own being in the world, as alive and therefore mortal. We all suffer because of the structural impossibility of completeness of satisfaction in a life marked by not knowing and loss, while constantly aware of and reaching for an important part of ourselves always elsewhere, insistent and powerful. That is our link to the dimension defining of the space of the unconscious, which remains outside the immediate reach of our three-dimensional embodied being and which can be accessed only by approximation, through language and articulation addressed to a particular kind of other: the analyst as the subject-supposed-to-know. This is the embodied Other who occupies a position that facilitates access to a personal space of desire whose structure encompasses both impossibility and change. The change that the analyst can bring is not that of resolving or denying the impossibility, but through facilitating transformations in how this particular landscape is navigated, so that more freedom and satisfaction can be found on the side of life.

It is important not to lose sight of the central point that human suffering can be understood as a structural impossibility brought about by the tension between being alive and aware of mortality at the same time. This is overlaid with the spatial dimensional discord (akin to the grammatical notion of lack of accord) between the three-dimensional space occupied by the body and the four-dimensional structure of the unconscious. Analysis works through the facilitation of operations on the structure of this space, leading to changes in the possibilities of movement within.

A critical feature of change in topological spaces is that local change can lead to global transformation. A change that modifies the structure of the space, even if introduced in a particular location, will have the effect of impacting on every point in that space. This is crucial to understanding how analysis, rather than the analyst, works. Even though analysands arrive with several complaints, and often say that they do not know where to start, the invitation of analysis is to start anywhere, with whatever comes to mind. To say something. Equally, what experience shows, is that once change begins to occur, it quickly unfolds in all aspects of life, bringing unanticipated transformation in areas that were not even talked about directly over the duration of the analysis (one analysand starts to paint, another is no longer constipated etc.).

Using Lacan's *La troisième* (2011) as the starting point, Bursztein (2017) sums up the essence of the cure as the combination of repeated cuts within jouissance and the elaboration of the fundamental phantasy of the subject at the level of the Borromean knot (p.27). Cuts are precisely the kind of topological transformations capable of bringing about a modification of the structure of the entire space where they are applied. Recall that jouissance transcends into the beyond of language and of the pleasure principle, and that the signifier provides the link and means of intervention between the space of the body and that of language. In Bursztein's formulation, words, 'access to the jouissance of the body marked by the lack constitutes the invariant finality of all psychoanalytical cures' (Bursztein, 2017, p.30). Interestingly, in topology invariants are properties that are preserved in the mapping between homeomorphic spaces (Alexandroff, 1961, p.7). In that sense, one can understand something about the value and possibility of each clinical encounter: what works is analysis itself, not the analyst. Through a variety and multitude of cuts and changes to pathways, transformation is ultimately possible not because of a particular analyst, but because analysis itself engages with and operates upon the unconscious as space. The Other, as embodied by the analyst, operates through language to effect changes in the space that escapes language. As Verhaeghe puts it, beyond signifiers, 'we meet with something different, where the signifier is lacking and the Real insists […] talking is not enough […]' (Verhaeghe, 2018, p.250). The interventions of the analyst, in Lacanian practice at least, go beyond speech, encompassing any intervention that amounts to a psychoanalytic act, that is to say an expression of the analyst desire guided by the ethics of moving the analysand towards the end of their analysis. What the analyst directs is the cure, not the analysand. All interventions, whether in language or not, operate upon the space beyond the reach of language.

Chapter 9 explores a reading of subjective psychic structures in spatial terms, while Chapter 10 offers a clinical illustration that brings to life the more abstract mathematical and psychoanalytic concepts at play. Chapter 11 concludes.

References

Alexandroff, P. (1961). *Elementary concepts of topology*. New York: Dover Publications.
Bursztein, J-G. (2017). *L'Inconscient, son espace-temps: Aristote, Lacan, Poincaré*. Paris: Hermann.
Evans, D. (1996). *An introductory dictionary of Lacanian psychoanalysis*. London: Routledge.
Lacan, J. (2020). *L'identification: Séminaire 1961-1962*. Éditions de l'Association Lacanienne Internationale. Paris: Publication hors commerce.
Verhaeghe, P. (2018). Position of the unconscious. In Vanheule, S., Hook, D., and Neil, C. (eds.). *Reading Lacan's Écrits: From 'Signification of the phallus' to 'Metaphor of the subject'*. London: Routledge. pp.224–258.

Chapter 9

The spatial unconscious and the clinic of psychic structures

There is much to be gained by thinking of the subjective experience of the analysand in spatial terms, both in relation to dimensionality and with regard to structural impossibility.

In an overview of Imre Hermann's contribution to understanding psychological processes spatially, Klaniczay (2007) stresses the creation of the most vast space possible as the fundamental question of therapeutic work (p.38). It may be more accurate to say that the space in question is not so much created as made accessible in new ways, as it exists structurally prior to any analytic intervention, but the fixity of symptoms leaves much of it out of reach. Speech in the clinic modifies and augments the subject's range of movement psychically in such a way that new life choices and relating options become accessible.

The spatial view of the unconscious explored here is consistent with both the spatial formulation attempts made by Freud and the Lacanian view of the unconscious structured like a language. Moreover, this particular spatial reading also makes it possible to understand why analysis works. If language is indeed 'in a timeless synchrony, whilst speech functions in a linear diachrony' (Verhaeghe, 2018, p.231), then language has the quality of space, with speech as movement in it. When two spaces are in some relationship of equivalence, changes in one of them can be mapped as changes in the other. However, for some pairs of spaces no one-to-one mapping is possible. Thus, there is no direct mapping between the unconscious and consciousness; no homeomorphism between these two spaces exists. Equally, language fails structurally, and truth can only be half-said. There is always a residual which escapes formalisation and remains in the Real, and a direct translation of experience into language is not possible; words can only operate as concatenation (Charraud, 1997, p.67).

Each person arrives to analysis with speech and a story half-told and half-sayable. The saying clarifies something of the consistency[1] of their particular space of language, while also revealing their particular way of circumscribing the impossibility that their being is organised around. Continuity cannot be built out of a series of discontinuous elements. This does not work in physics (Balibar, 2003, p.15), or in topology or in psychoanalysis, as the two states are fundamentally incompatible. Indeed, analysis does not aim at producing comfort and order, or at promoting

DOI: 10.4324/9781003479284-13

The spatial unconscious and the clinic of psychic structures   157

continuity where none could exist but at facilitating a subjective invention that deals with the fundamental unruliness and discontinuity of human experience. In other words, the work of analysis is undertaken as a working through of the experience of inaccessibility, so as to circumscribe and liberate something about the underlying impossibility.

The most complex formulation of subjective experience in spatial terms is Lacan's conception of the interlinking of three registers of experience in what he calls a Borromean knot. In a *mise-à-plat* representation, each register operates as some kind of boundary centred on the space allocated to the object cause of desire, *obj a*, and demarcating a particular type of jouissance.

It is easy to regard this structure as given, yet this is not something ready-made, but rather comes about as the outcome of a process whereby, through language, an infinite space of possibility becomes closed into something manageable, the subjective space. This is what Bursztein (2017b) refers to when he clarifies the development of psychic structures in terms of the presence of compactification (in neurosis) and its absence in psychosis, and also in terms of the location of what he calls 'the fourth consistency' (p.71). The possibility of compactification depends upon the presence of maternal and paternal conditions which allow for an actualisation of the Borromean potentialities of the structure (R, S, I) (Bursztein, 2017b, p.80). Here, compactification refers to the closing of the infinitely open line into a circle or ring. The underlying Oedipal configuration may or may not allow for the deployment of the phallic function, that is to say, of the point at infinity in terms of the spatial structure. In this sense, psychic structures can be thought of as topologically distinct configurations. Bursztein (2017b) proposes a direct correspondence between the failure of the paternal function in the task of compactification (and therefore creation of subjective space) and the emergence of particular psychic structures.[2] Thus, psychosis is a failure of unfolding of the paternal function in the Real; paranoia follows from such a failure in the Symbolic and neurosis emerges from this failure in the Imaginary (p.41). As we have seen in Chapter 6, each structure has a particular relationship to impossibility, which indicates that, in each clinical encounter, we are invited to grasp something about the topology at play.

The spatial complexity of this formulation of subjective experience can be situated in a nested progression of psychic dimensionality, which moves from the one-dimensional space of identification, through the two-dimensional space of the clinic (Lacan's L-schema), to the three-dimensional domain of the interplay between the Imaginary, the Symbolic and the Real (the Borromean knot), all the way to the four-dimensional domain of psychic structures as captured by Lacan's R-schema. It is this sequence that the remainder of this chapter considers.

9.1   The recurrence of suffering

A three-dimensional sphere encountering a plane would appear, from the vantage point of the receiving space, as a circle, or as a sequence of circles of variable diameters, if the sphere moved in a direction external to the plane. Similarly, a four-dimensional sphere would be perceived as a sphere in three dimensions, or as a sequence of spheres of varying diameters, if in movement in the direction of the fourth dimension. This sequence of spheres is in fact a sequence of three-dimensional sections through the four-dimensional sphere. What comes across as a sequence of similar and distinct objects is, in fact, a truncated and discontinuous perception of a whole entity which, in a space of lower dimensionality, cannot be perceived in its totality as one.

If we translate this understanding to transference or the repetition compulsion, we become able to consider how the recurrence of suffering is not a case of discrete repetitions, but of incomplete representations of a whole that cannot be perceived in its totality. Our perception of connection and interdependence can, then, to some extent, be the outcome of the limitations in perception of a distinct and continuous totality that is not immediately accessible. A limitation in the perception becomes transformed in a view of discontinuity that is attributed to what is being perceived rather than be understood as a limitation on the part of the *percipiens*, or perceiving subject. This is the lure of the Imaginary, with its illusion of completeness, in contrast with the disruption of discontinuity of the Real, such as it is first grasped by the emerging subject at the mirror stage. The image which promises completeness and unification becomes an ideal which operates as a defence against the experience of felt fragmentation in the Real body one *has* (not *is*). The subject is endlessly caught between experiencing as fragmented that which is continuous elsewhere, while trying to bring continuity to a fracture which cannot be repaired, to a structural discontinuity.

Occasionally one encounters analysands whose presentation is a suffering directly and openly linked to the question of life and death. An interesting solution that sometimes emerges for such subjects is not to think of death as discontinuity, but of their very existence, which they come to regard as a hiatus between their non-existence prior to conception and their non-existence into death. For many this is pacifying enough to make being and feeling alive possible.

In his end of century mathematico-philosophical musings, Hinton, a contemporary of Freud's, considers that, if a fourth dimension exists, either we have a three-dimensional existence only or we have a four-dimensional existence, but we are not conscious of it (Hinton, 1884, p.9). I propose the latter, with the specification that we are unconscious of it, which is not the same as non-conscious. This is not an intellectual proposition, but one rooted in the experience of dreams. In particular, I am referring to that quality of dreams whereby things not normally found in proximity in waking life (because of either spatial or temporal distance) can be experienced as close by and simultaneous in dreams. The relevance of an extra dimension can be illustrated with a simple example: shapes or traces at opposite ends of a two-dimensional space (plane) which cannot be brought in continuity with each other in the plane can achieve proximity if the plane is either folded or rolled, using the possibility of movement in an additional dimension, away from the plane itself, that is to say, in three dimensions. Figure 9.1 illustrates this, with the two half disks on either side of a strip brought into continuity with each other by turning and folding the two-dimensional space into a three-dimensional structure.

The spatial unconscious and the clinic of psychic structures 159

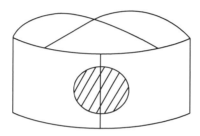

Figure 9.1 Continuity in a higher dimension.

This is consistent with Lacan's view that dreams are not the unconscious, but something of the unconscious is transcribed in dreams, something defined by a topological structure (Charraud, 1997, p.62).

Something particular to topological spaces pertains to the relationship between the local and the global. A common mathematical way of describing such a relationship is in terms of fractals as repetitions at different scales, making everything appear both the same and different. This is a particular kind of repetition and Freud describes such a structure without naming it in the context of his very early work, in his *Project for a Scientific Psychology* of 1895, where he is still very much expressing his ideas in neurological terms: 'A single neuron is thus a model of the whole nervous system' (Freud, 1950, p.298). In more familiar psychoanalytic language, Lacan also seems to hint at something of the sort as early as Seminar II: 'This dialectic is found in experience at every level of the structuration of the human ego' (Lacan, 1991, p.50).

In Lacanian terms, repetition is essential to the definition of the unconscious, and it is primarily understood on the side of the intermittent opening of the unconscious (Miller, 2008, p.11). As we have seen in Chapter 5, Lacan make a distinction between repetition that pertains to the encounter with the Real (*tuché*) and repetition that is in the nature of the insistence of the Symbolic order (*automaton*)

(Lacan, 1979, pp.53–55). This repetition pertains to trauma understood as an encounter with discontinuity, expressed by both Freud and Lacan in spatial terms. Furthermore, this understanding raises the question of the lack of homeomorphic correspondence between registers of experience. The links in the Borromean knot cannot be conceived of as homogenous, despite the symmetrical representation used. The dimensionality of the Real surpasses that of the Symbolic. What is at stake is a gap between domains that cannot be engaged with other than by making a jump. This is the gap between the conscious and the unconscious or between primary and secondary processes, akin to the gap between spaces of different dimensionalities.

9.2 Identification and subjectivity

Armstrong (1983) spells out the essence of a very simple idea for constructing new topological spaces. As we have seen in Chapter 7, where we examined the construction of key topological structures, very different things happen when we start from a rectangle and 'zip up' opposite sides with or without a twist. A new space is produced by starting with a simple space and identifying some of its points. Note that here identification is used in the mathematical sense, which is that of making pairs of points identical to each other, reducing two to one. He calls the outcome of this process an identification space (Armstrong, 1983, p.65).

Although not explicitly formulated thus by Lacan, the unconscious as a four-dimensional space is fully consistent with his understanding of the relationship between the subject and the Other, in particular in terms of the clinic. While the IPA tradition is grounded in the view that the analyst can interpret the transference from a privileged external position with regard to the patient, in Lacanian analysis, both patient and analyst are *in* the transference. In the same way that we cannot apprehend physically a fourth dimension in a three-dimensional space, in other words in the way that we cannot apprehend the nature of the space we inhabit, neither can the analyst be outside the transferential space they occupy in relationship to their patients.

The concept of projection has long been used in psychoanalysis, to denote a disowning of affect by the subject and an attribution of this to others around them. This is often paired with the concept of introjection, which is seen as an avenue for identity formation, primarily in analyses aligned with the IPA tradition, where, in essence, some identification of the patient with the analyst is the desired outcome of the cure (Sandler and Dreher, 1996).

Mathematically, projection has two important consequences of psychoanalytic relevance. The first is the loss of dimensionality it entails, combined with the impossibilities this creates. The physical equivalent is that of a loss of a degree of freedom, and hence of impaired movement. The second is the element of distortion that projection introduces, by the confounding and overlapping it generates, as, for instance, the impact of transferring the details of the spherical planet to the plane of a map. This distortion is one of expansion or shrinking, a kind of stretching that preserves something about the contours but not the relationship between the points contained by them.

Our embodied relationship to spatial experience, as represented in Lacan's optical schema in Seminar I, highlights the importance of the subjective position. In that sense, space is not homogeneous, and where the subject is located defines the outcome. The work of analysis is towards the unfolding of the most singular expression of the subject as position in their psychic space. This is very different from the prevailing IPA view whereby the outcome of analysis is a uniquely determined point of identification with the analyst.

The critique of collapsed dimensionality has been elaborated by Lacan in relation to the IPA approach in two main ways, at the level of technique, and, fundamentally, in terms of direction of treatment. Clinically, one encounters the ongoing struggle with one's own memories, not just in terms of the building of a history or the construction of a personal recollection of the past, in a way that generates a satisfying (and therefore deceiving) narrative, but most of all with the odd fragments of memory, little disconnected shards, traces that do not seem to lead anywhere in the domain of 'sense', but persist. Such discontinuous occurrences are also in the nature of dreams: a word, a sensation, a flicker of something, a recurring invitation to decode. To relate to Freud's archaeological references, the shards I am referring to would not make it easy to reconstruct the pot they came from (although one often finds in museums vessels where there is much more reconstruction material than original remains). The thing that can easily escape is that, for that pot, the remainder of it is elsewhere. It is the same with the unconscious, something inaccessible is 'in abundance elsewhere'.[3] In spatial terms, the relationship between the fragmented known and the continuum of the unknown is well illustrated by the way a crumpled and then un-crumpled, wrinkled piece of paper sits on a flat surface: it touches it at a few points which appear disparate and nonsensical. The rest of the crumpled space, in its continuum, floats above the flat surface, there all the while, continuous with itself, but in another dimension, as far as the surface it sits in is concerned.

9.3 Dimensions in the clinic, revisited

Lacan's critique of the IPA/Kleinian approach addresses both the shortcomings in technique and the incompleteness of theoretical formulation. In essence, an analytic encounter centred on the patient-analyst axis, through transference and counter-transference interpretations, keeps the work in a unidimensional space, that of the Imaginary (the ego-Other axis). In his L-schema, Lacan distinguishes between the symbolic and imaginary axes: the recognition of the new dimension of the Symbolic as distinct opens up a new space with increased possibilities for movement and stability. Analysis, in this sense, is a process that introduces a new dimension in relating, by separating the hitherto confounded domains of the Imaginary and the Symbolic. This unfolding of collapsed dimensions recognises a new space in which the subject can emerge.

Following on from the two-dimensional L-schema of his early work, Lacan develops a four-dimensional spatial articulation of the relation between the subject, the Other and desire, in what is known as the R-schema (Lacan, 1956, p.462). In this new formulation, the Oedipal structure is added explicitly to the structure of the Imaginary-Symbolic dynamic (Leader, 2003, p.187). Crucially, this is done with explicit consideration for the spatial configuration that encapsulates the relationships between the elements concerned. Specifically, Lacan places this structure on a cross-cap and specifies the topological nature of this space in a footnote added ten years later.

The R-schema in *mise-à-plat* can be thought of as a two-dimensional projection of a tetrahedron. If we consider the subject, the mother, the father and the phallus as vertices of such a three-dimensional construction, it is possible to recognise that each vertex is in a different plane from the one defined by the other three. The number of vertices remains the same, but in this two-dimensional representation it becomes possible to recognise something more about the nature of their interrelationship. In the words of Blum and Secor (2011),

> [T]he R schema demonstrates Lacan's quadripartite structure of the subject; the symbolic, Oedipal triangle (I-M-F) is redoubled by an imaginary triangle (e- j -i), and these two configurations maintain and are maintained by a split in the field, the R(eal) quadrilateral.
>
> (p.1039)

At the same time, in the fundamental polygon representation which Lacan uses to give spatial consistence to this schema, the four corners of the rectangle are, in fact, only two distinct vertices, one for each opposing pair of corners. The space appears flat, but contains two twists which interplay with each other in a way that defies representation in three-dimensional space. Notice that diagonally across this representation there is the space of a Möbius band delineating the field of the Real which cuts across the Symbolic, a hole that escapes representation and around which the entire structure is arranged.

In a close reading of Lacan's paper that introduces this schema, Vanheule (2020) allocates no consideration for its topological implications, yet he recognises the fundamental structural discontinuity at play, stressing its relevance for the clinic, where the analyst is invited to focus on 'the discontinuities in the patient's speech [...], rather than hidden meanings associated with symbols' (p.181).

Miller (1982) underlines the change in Lacan's thinking on the notion of phantasy, from its early placing in the Imaginary register to the later recognition of its role in tempering jouissance and the inertia of the Real, which he comes to define as the impossible, that which remains outside symbolisation (Cléro, 2002). Lacan shifts emphasis away from the supremacy of the Symbolic in his early work, towards a heterogenous yet non-hierarchical Borromean structure, in which the Real is 'that which always comes back to the same place' and where it cannot be met by the thinking subject (Lacan, 1979, p.49).

What phantasy veils is the hole, the impossibility of the absolute and non-existent jouissance perceived as the lost or forbidden enjoyment of incest. It does this in an attempt to offer an answer to the subject's encounter with lack of being, by weaving its object into the imaginary of an autoerotic object (Bursztein, 2017b, p.77). Love takes the place of fear as remedy for mortality.

The movement along the boundary that circumscribes the hole is an iterative sequence, from lack to phantasy and back, which is constantly re-written around moments outside time, producing a structuring effect of the type before-after, while operating like limit points in mathematics (Bursztein, 2017a, pp.35–36). In topology, limit points are those that can be approximated by points in their neighbourhood. Lack is a point of discontinuity, while phantasy gives an approximation of continuity.

This is central to the role of psychoanalysis in the clinic, where the aim is to move the patients towards perceiving their own fundamental phantasy and, beyond that, to replacing this with the play of a series of phantasies of desire (Bursztein, 2017a, p.49). In terms of the boundary around the *obj a* region, the accompanying movement is away from a thick boundary on the Imaginary to a thick boundary on the Symbolic (Bursztein, 2017a, pp.52–53).

Phantasy can be thought about as a point of singularity in three dimensions, a point of spatial discontinuity wherefrom a new dimension emerges in the process of analysis, making possible movement in a direction that is radically away from everything that can be perceived in the space of the body. The traversing of the phantasy that Lacan speaks of is an unequivocally spatial formulation of the transmutation that the testimonies of the pass in the Lacanian tradition speak to (Miller, 2022). A domain of possibilities previously unavailable becomes conceivable and then accessible through the work of analysis. In other words, another space opens up, which is radically different, not larger, further away or more of the same. This is not arrived at through knowledge, but through disruption, and the process requires the analyst to occupy the position of *obj a*, located as a limit or vanishing point, or in the position of the point at infinity in the spatiality of the cross-cap structure. The analyst remains in the function of the hole, outside the consistency of the plane but with a structuring effect on it.

Movement within the space of psychic possibilities requires particular attention to perceived points of discontinuity, which open up into a higher dimensionality, a realm accessible incompletely but unequivocally through the mechanism of language.

Notes

1 Namely, that which surrounds the hole and coexists in a mutually defined way with it.
2 It is interesting to see Bursztein does not recognise perversion as a structure, while distinguishing paranoia from psychosis (presumably schizophrenia).
3 I owe this powerful expression to artist Lara Geary.

References

Armstrong, M.A. (1983). *Basic topology*. Springer-Verlag: New York.
Balibar, F. (2003). Le réel a toujours eu quatre dimensions. In Cohen-Tannoudji, G. and Noël, È. (eds.). *Le réel et ses dimensions*. Editeur: EDP Sciences, pp.11–23.
Blum, V. and Secor, A. (2011). Psychotopologies: Closing the circuit between psychic and material space. *Environment and Planning D: Society and Space*, 29, pp.1030–1047.
Bursztein, J.-G. (2017a). *L'Inconscient, son espace-temps: Aristote, Lacan, Poincaré*. Paris: Hermann.
Bursztein, J.-G. (2017b). *Subjective topology: A lexicon*. Paris: Hermann.
Charraud, N. (1997). *Lacan et les mathématiques*. Paris: Anthropps.
Cléro, J.-P. (2002). *Le vocabulaire de Lacan*. Paris: Ellipses.
Freud, S. (1950 [1895]). Project for a scientific psychology. SE1, pp.281–397.
Hinton, C. (1884 [1980 reprint]). What is the fourth dimension? In *Scientific Romances, vol.1*, Speculations on the fourth dimension. Selected writings of Charles H. Hinton, New York: Dover Publications.
Klaniczay, S. (2007). Espace et psyché: À la mémoire de Imre Hermann. *Le Coq-héron*, 188, pp.35–41.
Lacan, J. (1956). On a question prior to any possible treatment of psychosis. In Fink, B. (ed.). (2002). *Écrits*. London: WW Norton, pp.531–488.
Lacan, J. (1979[1973]). *The four fundamental concepts of psychoanalysis*. London: Penguin Books.
Lacan, J. (1991[1978]). *The seminar. Book II. The ego in Freud's theory and in the technique of psychoanalysis, 1954–1955*. London: WW Norton.
Leader, D. (2003). The Schema L. In Burgoyne, B. (ed.). *Drawing the soul: Schemas and models in psychoanalysis*. London: Routledge, pp.172–189.
Miller, J.-A. (1982). From symptom to fantasy and back. The symptom 14. https://www.lacan.com/symptom14/from-symptom.htmlhttps://www.lacan.com/symptom14/from-symptom.html, accessed 11 March 2021.
Miller, J.-A. (2008). Transference, repetition and the sexual real. *Psychoanalytical Notebooks*, 22, pp.7–18.
Miller, J.-A. (2022). *Comment finissent les analyses: Paradoxes de la passe*. Paris: Navarin Éditeur.
Sandler, J. and Dreher, A.U. (1996). *What do psychoanalysts want? The problem of aims in psychoanalytic therapy*. London: Routledge.
Vanheule, S. (2020). On a question prior to any possible treatment of psychosis. In Hook, D., Neill, C. and Vanheule, S. (eds.). *Reading Lacan's Écrits: From 'The Freudian thing' to 'Remarks on Daniel Lagache'*. London: Routledge, pp.163–205.
Verhaeghe, P. (2018). Position of the unconscious. In Vanheule, S., Hook, D. and Neil, C. (eds.). *Reading Lacan's Écrits: From 'Signification of the phallus' to 'Metaphor of the subject'*. London: Routledge, pp.224–258.

Chapter 10

Clinical implications

My recommendation in the introduction was for the reader to follow the text, chapter by chapter, in the order presented. It is in the nature of mathematics that understanding builds gradually and that it pays to follow through all the steps in order to master the mechanisms at play. From the point of view of the clinician, this may be the most important chapter, but none of it can be understood in the way intended without a grasp of all the notions examined thus far.

It is now possible to see how it all comes together in the clinic, and to consider how the interplay between all the technical notions explored in the preceding chapters illuminates the effects of psychoanalysis in producing structural changes in the psychic space by opening up access to new dimensions and by circumscribing impossibility in spatial terms. The particular blend of mathematical and psychoanalytic notions explored up to this point can now present its relevance to understanding in a new way the underpinning mechanisms of change in the analytic process, and to locating the role and work of the analyst in the unfolding of this unique, subtle and powerful process.

10.1 Topology in the clinic

There is no consensus or clarity over the use of topology in the psychoanalytic clinic (Hughes, 2013). For many, the essence of topology remains largely unknown and implicitly irrelevant, while for some it offers unique access to addressing something about the essence of the human experience. Burgoyne, for instance, goes as far as to declare that Lacan's seminar on Identification 'contains a mathematics of the human soul' (Burgoyne, 2018, p.18).

To a large extent this lack of integration is a consequence of the fragmented nature of the profession and of the formation upon which it rests. It is as easy to become an IPA analyst without ever encountering Lacan's ideas as it is to practise as a Lacanian analyst without any familiarity with the work of Matte Blanco or Bion. Even among Lacanian analysts, engagement with topology remains a matter of personal preference and inclination, rather than a considered position. It is, nevertheless, important to consider what is at stake, if we are to take seriously the

DOI: 10.4324/9781003479284-14

implications of this entire body of knowledge for the effectiveness of our work in a clinical setting. So, let us recap.

Lacan's use of mathematics, like his use of language, changes fundamentally the relationship to what can be known and how. Following a trajectory that echoes the development of the exploration of both mind and space, he moves away from a view of infinity as something pertaining to magnitude, towards emphasising the inbuilt impossibilities in the landscapes of the mind. Centred on the Imaginary register of experience, the prevailing IPA analytic literature which he critiques, explores the relationship between internal and external, introjection and projection, container and contained; in other words, it inhabits a Euclidean world. On the contrary, Lacan uses the Symbolic register to move away from such binary partitions towards topological representations of the human experience. His emphasis is on dis/continuity and transformation, and not on classification and dichotomies.

Charraud (2004) examines the radical change introduced by Lacan, away from spatial structures organised around a centre, in the way that a sphere is, towards topological objects whose properties are not just similar to those of the psyche, but constitute its very structure. She expands on the ensemble of topological spaces used by Lacan as structures between which there is the possibility of movement and stresses how, with the Möbius band and the cross-cap, it is possible to understand how that which is most internal is also most external (Charraud, 2004, p.137). The recognition of this one-sidedness (non-orientability) of psychic space is also what resolves something that had puzzled Freud greatly: namely, the possibility of a double inscription in different agencies of the psyche. Thinking of mental space as one-sided in this manner explains the possibility of 'passing from one side to the other' without a break. As Charraud stresses, the logic of the subject of the unconscious is primarily Möbian. What is locally two-sided is globally one-sided.

In his earlier work, Lacan conceives of the unconscious as a locus, which he calls the Other. This is neither a person, nor a place. Thus, he makes use of the Euclidean geometric concept of locus in order to convey the constitution of the unconscious as a function rather than a place that can be mapped topographically (Lacan, 1997). In his later elaborations, he moves beyond Euclidean conceptions to an articulation of space that is less constraining as much as it is more nuanced. Even though he does not express this in terms of dimensionality, Lacan hints at a higher dimension when he articulates the limitations of expecting to be able to place the unconscious in the world we can see: 'when the Other with a big O speaks it is not purely and simply the reality in front of you [...]. The Other is beyond that reality' (Lacan, 1997, pp.50–51); it is 'the Other one addresses oneself to beyond what one sees' (Lacan, 1997, p.56). To Lacan, knowledge is synchronic and topological, defined by proximity and neighbourhood, and in Seminar XX he sets himself the task to demonstrate a strict equivalence between topology and structure (Lacan, 1998).

Lacan links explicitly algebraic topology to what he is trying to offer on the symbolic plane: a logic that is elastic and flexible (Lacan, 2020, p.230). With his move into topological thinking, psychoanalysis becomes explicitly spatial, with topology showing 'the real of structure which cannot speak itself' (Ragland, 2002, p.122), and presenting the foundations of the position of the subject. In Greenshield's apt summary, 'topology allowed Lacan to present and demonstrate the structural paradoxes that define the psychoanalytic subject as distinct from the subject of conscious self-apprehension' (Greenshields, 2017, pp.32–33).

Central to Lacan's entire pursuit is the notion that a subject is created by language. Topologically, he expresses this in striking terms in Seminar XI, where he states that 'it is around this signifier of the cut that what we call a surface becomes organised' (Lacan, 2020, p.388). It is the cut that makes the surface, rather than the surface preceding the cut. The space of subjective experience which the surface constitutes emerges as a result of the cut. This resonates strongly with Freud's paper on negation (Freud, 1925), where existence becomes structured by a first mark. Lacan likens this process to the act of the potter who creates a vase around a hole, forming a substance around an emptiness which 'does not pre-exist the arrival of the substance' (Greenshields, 2017, p.41).

The spaces of subjective experience and those of language are both conceptualised by Lacan in topological terms. Language itself amounts to a topological space such that the pathways of movement on the surface in question are those traced by chains of signifiers. New pathways can be constructed such that both the hole and the consistency can be traced. Patients arrive with a lot of empty speech, at the core of which there is a question waiting to be articulated in full speech. Their stories of something that repeats itself, of feeling stuck in the constraints of their destiny, circle the void of desire in the recounting of unmet demands.

Clinically, it is important to consider how interpretation can operate as a way of altering these pathways of speech, by cutting the chains of signifiers and opening up the possibility for new connections to form and for new pathways to emerge, such that demand and desire can be repositioned with regard to each other in a continuity that recognises rather than avoids the central impossibility at the heart of and defining of the structure. In order to be able to operate through intervention, the analyst needs to have some systematic understanding of the structure of the space in question and of the possibilities and impossibilities of movement within it. This is why a grasp of spatial dimensionality and of topology matters, if one is not to operate blindly.

One immediate way in which all of this comes into play in the clinic is in terms of understanding the ways in which something can change at the level of repetition in suffering. Is it enough for the patient or analysand to be given a point of address, or does something else need to happen so that repetition can be touched and something new can emerge? We can answer with some grounded clarity if we think topologically about interpretation as an operation upon a chain of signifiers, which in turn can be understood as a pathway on a topological surface.

In the 1961–1962 Seminar IX on *Identification*, Lacan examines the torus as structure of the subject, expanding on his early reference to this topology in 1953, in *The Function and Field of Speech and Language in Psychoanalysis*. His focus is on the topological properties of the torus, as contrasted with those of the plane or sphere. Specifically, he concentrates on possible trajectories on the surface of a torus, distinguishing between those that can be reduced to a point or collapsed to a tautology (Lacan, 2020, p.187) and those that constitute full circles, the irreducible pathways. The pathways that loop around the consistency 'materialise the metonymical object of demand' (Lacan, 2020, p.207), whilst those moving around the central hole circumscribe desire. In the original text, these pathways are called '*lacs circulaires*' (e.g. Lacan, 2020, p.232), meaning paths or loops, and not circular axis, as in the unauthorised Gallagher translation. The circles of demand are those of the circle generating the torus by revolution, while the circle of desire is the path of this revolution.[1] In this sense, it is possible to grasp the relation between demand and desire as captured in metonymy, that is to say, in the articulation of desire as a succession of repetitive demands. Lacan wonders 'to what extent can this shape allow us to symbolise as such the constituents of desire, inasmuch as the desire, for the subject, is that something which s/he constitutes on the path of the demand' (Lacan, 2020, p.236). The circles that cannot be reduced to a point show, in his words, that the torus is not a mere puff of air, but has 'all the resistance of something real' (Lacan, 2020, p.190).

A particular 'privileged loop' combines demand and desire, and circumscribes the location of *obj a* and of the phallus in relation to the void circled by the torus (Lacan, 2020, p.237). With regard to this void, Lacan spells out the relation of the subject to 'nothing' in terms of destructive nothing (aggression), the death drive and dialectical Hegelian negativity (Lacan, 2020, p.241), specifying that the nothing of the subject is 'nothing as such' (*ibid.*). The void is constitutional of the torus, circumscribing the nothing which operates as a kind of placeholder for the object of desire: 'The object is not fixed or determined here by anything else other than the place of a nothing which, we could say, prefigures its eventual place, without allowing us in to locate it in any way' (Lacan, 2020, p.390). This resonates with the function of zero not as a number as such, but as a placeholder, a point where nothing in particular is located, but from which (spatial) dimensions can be unfolded. If we think in terms of a system of axes, zero is the only point in common to the dimensions represented by each axis. Furthermore, 'The object of desire is only constituted in relation to the Other. There is nothing special about the object other than this absurd value given to every trait as privileged' (Lacan, 2020, p.242). Thus, desire is constituted on the path of a question of birth/non-being [*naître/n'être*] (Lacan, 2020, p.248).

The culmination of Lacan's exploration of the relation between nothing and void is captured in the interlinked tori he examines, in the way they embody the interplay between the desire of the subject and that of the Other. The consistency of demand of one 'fills' the gap of the other's desire. The subject and the other are 'two tori linked by means of each other's lack' (Greenshields, 2017, p.67; Figure 10.1).

As Lacan clarifies, in the case of neurosis, one tries to place in every demand the object of one's desire, attempting thus to obtain from the Other the satisfaction of this desire, 'that is to say precisely what cannot be demanded' (Lacan, 2020, p.248). Substituting demand for desire leads to this interlinking of tori, a kind of dialectical knotting. The obsessional neurotic desires the Other's demand, while

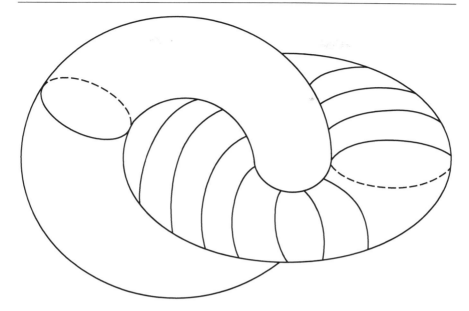

Figure 10.1 Interlinked tori: desire and demand.

in hysteria, the subject demands the Other's desire, presenting him- or herself as the enigmatic cause of this desire (Greenshields, 2017, p.68). In Lacan's rereading of Freud, 'desire is structured by the knot called Oedipus' (Lacan, 2020, p.215), namely, the relationship between a demand that becomes law and the desire of the Other. The law is that 'you cannot desire what I have desired', with the subject's path of desire having to include this hole of the exclusion. Desire is constituted 'in the original interdiction' (Lacan, 2020, p.224). Nevertheless, this prohibition of the law is nothing more than a veil to the impossibility of satisfaction, of that which lies beyond the pleasure principle. The hole around which this dance takes place is structural, not an outcome of the law.

It is also in this seminar on *Identification* that Lacan introduces the cross-cap/projective plane as the structure of the fundamental phantasy. As we have seen, the cross-cap is an instance of a projective plane, a closed (i.e. without a boundary) topological surface also known as a twisted sphere or a sphere with a hole (Volkert, 2013, p.1). It is a plane together with a line at infinity (also called an improper line, an infinitely distant one – Seifert and Trelfall, 1980, p.10). This is the higher-dimensionality equivalent of a point at infinity where all lines meet, including both ends of one line. The line is to the plane as the point is to the line, so a projective plane 'contains' a single line at infinity at which the edges of the plane itself meet.

This line is neither nothing nor a void, although if one tries to make a start on building a cross-cap, one quickly sees that it is the very starting point that remains somehow outside the entire construction which begins to take shape around it. This point is there by not being there. It is the only point which does not connect to any other points, in the way that the phallus is the only signifier which can signify itself, but which cannot be named (Lacan, 2020, p.335).

Lacan describes the cross-cap as a structure 'particularly suitable to show us the workings of the incessant element of desire as such, in other words of the lack' (Lacan, 2020, p.365). Within this structure, the point that I was referring to is the point 'by which we symbolise that which can introduce any *obj a* in the place of the hole. This privileged point, we know its functions and nature: it is the phallus, the phallus inasmuch as it is through it, as operator, that an *obj a* can be put in the very place of which we can only grasp the contour of, in the torus' (Lacan, 2020, p.366).

In other words, the line at infinity is where a cross-cap immersed in three-dimensional space appears to self-intersect, and the location of this apparent self-intersection is where Lacan, precisely, recognises the place of the phallus. He calls this a point-hole, *point-trou* (Lacan, 2020, p.421), and regards the cross-cap/projective plane as a way of organising this particular hole. This is the point around which turn the two coils of the cut of the interior-eight (Nasio, 2010, p.95). Structurally, the point-hole is the spatial expression of the impossibility of reducing the Möbius band to a point. Structurally, its edge 'circles' this hole twice over, along its double-looped edge. It is not apparent in either the Möbius band or the disk, but exists as a kind of remainder in the imperfect operation of joining them up. However, Nasio (2010) argues that, after the cut, this point remains with the disk (*obj a*), marking it with its signifying value as object of desire. It is crucial, however, to remember that such a point is not intrinsic to the theoretical surface, but is the result of its immersion in three-dimensional space, which is where the body pursuing its satisfaction resides.

This links to Lacan's notion of extimacy, whereby 'that what is *most* interior is most exterior and vice versa' (Greenshields, 2017, pp.89–90). Locating jouissance at a potential point at infinity gives it a place that encompasses pleasure and pain, beyond the '*finitude* of signifying combinations' (Greenshields, 2017, p.90). The subject and jouissance are heterogeneous to each other, and this particular relationship is captured in the topological structure of the projective plane, a heterogeneous composite itself, combining a one-sided Möbius band with a two-sided disk, brought together alongside a three-dimensional circular boundary, the inverted eight. Miller (1979) stresses the importance of this heterogeneity, as a way of underlining that there is no direct correspondence or overlap between the subject and their jouissance.

Notice how the projective plane structures the space of the dialectic between self and Other, with and around the point at infinity which is neither in oneself nor in the Other, who is also lacking. It is not here not because it is elsewhere. Its non-existence can only be marked by adding this point at infinity, by positing it. This is important because doing so makes possible some operations (movements) otherwise not available. This is similar to the effect of the introduction of imaginary numbers, but is more specifically related to the question of points of discontinuity in space. Once a missing point is acknowledged by positing its position at infinity, the space becomes compactified and thus open to navigation, with movement defined rather than blocked by impossibility. In the words of Nasio, the unconscious can be thought of as structured, and this 'must be imagined as a network bearing a hole' (Nasio, 1998, p.80).

10.2 Clinical illustration

Having considered in some detail numerous theoretical formulations pertaining to an understanding of the unconscious as space, it is time to turn to a clinical illustration.

Two main constraints need to be taken into account at this point: first, the limits imposed by the paramount value of confidentiality in any clinical encounter; second, the specific complexity of Lacanian clinical practice. Interestingly, both constraints converge on the point of singularity of a case. Confidentiality is commonly dealt with by modifying the details of a case to the point where no individual traits remain recognisable, although, on occasion, consent from a particular patient may authorise the publication of details. By modifying the details, however, one eliminates precisely the points of singularity, erasing those very traits that are specific to the individual in question and which are irreducible, hence obliterating largely the essence of what is to be transmitted. The second constraint arrives at the same point via the balancing of the underpinning theoretical considerations concerning psychic structure, and the radical uniqueness of each case, which cannot be reduced to a structure or anything else but itself. A further distinction exists here between particularity and singularity, with the former designating that which locates each subject in relation to a particular set of signifiers, and that of singularity, which operates like an objection to this set (Izkovich, 2022). A further complication is added here by the limitation inherent to using a specific example in order to illustrate a general hypothesis. This is only valid and possible on the understanding that the illustration is not used as proof, but as a way of showing how the proposed understanding of the unconscious as space can be put to use in the clinic.

With all this in mind, the focus of this illustration is on the specificity of the technique rather than on the particular profile of a specific subject. The vignette that supports this exposition is concerned with a scenario where the question of discontinuity arises both in the body and in language, and aims to establish some explicit links between Lacan's contributions to understanding language and the body as topological spaces, in the clinic of bodily discord. The more theoretical aspects relevant to this exploration with particular reference to the body are fully developed in Carrington (2020), on which this illustration draws.

Consider the case of an analysand who suffered a severe abdominal infection which led to a surgical intervention that transformed his body, both in appearance and in functioning, diverting one end of the colon to an outside stoma bag. The cut produced a hole. Although he appeared to manage with relative ease the adjustment to all the practical aspects of this major transformation, the analysand spent a great deal of time talking not about his body, but about the stoma bag, and all his strategies to ensure that he emptied it on time so that no further complications could arise. In other words, articulating something about the disruption at the level of the consistency was easy enough. That is to say, what was there could easily be spoken of in terms of continuity and avoiding of disruption. It was the discontinuity and the disruption that remained outside a direct address. In this sense, the stoma bag was, in speech and in the psyche, the placeholder for the hole; it marked its location while also veiling it. At the same time, the pathway circling the hole was virtually impossible to be traced directly and explicitly through language, as speaking directly about the body did not address the gap itself other than indirectly. In other words, the hole could not be talked about by merely talking about the consistency. Instead, the analysand became increasingly preoccupied with things not joining up in general, with various instances of gaps and discontinuities in his professional and personal relationships. The move that could not be produced in the Real, from the bag to the hole in the body, but from the veiling of the hole to other points of discontinuity, mapped only in the Symbolic. His concerns were not of a psychotic nature, but neurotic,[2] focusing on complaints around the lack of smooth rapport in his relationships, on failings and gaps that he resented feeling were always left to him to resolve, and about which he experienced a renewing sense of failure. It was this specific complaint around failing that occupied most of his time in analysis, rather than the disruption to his life caused by the surgical intervention, despite what the initial presentation might have indicated. Failure was a loaded signifier: the parents' marriage had failed; he had failed to keep them together and later on failed in every intimate relationship, feeling responsible for every single breakup. In his view, talking was destined to fail, too. His moments of impasse in analysis were always expressed in an almost identical formulation, often spoken in anger: *I am a failure, that is how I feel, that is never going to change, what is the point in talking*? The analysis lasted three years; there was a lot to say about impossibility, under the guise of failure. Also, despite recurring medical complications and follow-up hospital visits, the analysand missed only one session over the entire duration of his analysis, and kept the work going online or on the telephone on those occasions when he could not attend in person. No more gaps were to be created, with the Symbolic used as a means of sustaining continuity.

One particular detail from the analysand's early life held powerful relevance: he had been born prematurely and had spent the first few weeks of his life in an incubator. He knew nothing about the reasons for his premature birth, which was something never talked about in the family. As for his own experience of it, in such circumstances some form of memory can be understood to remain as mnemic traces, marking the subject's desire, phantasy, his 'very life force' (Vanier, 2015, p.150), the kind of traces that often appear as symptoms in adults who were born prematurely.

In terms of the body and its topological structure, starting life in an incubator, where '[T]he body orifices are deprived of their functions' (Vanier, 2015, p.161), where new orifices are created, and machines and implements are attached to these, means that it is most likely that the analysand had not been experiencing such invasions to his body as space for the first time as an adult. Indeed, this recent surgical experience was a second such encounter with bodily discontinuity, which came to establish something about the trauma of the first.

Most careful attention was needed to attend to all the emerging signifiers of discontinuity which referred to both consistency and hole, so that his position with regard to the new configuration of his body, which had been forced upon him, producing uncomfortable closeness to the knowledge of a gap at the level of his being, could become sustainable. The bodily healing process was slow, but he returned to work and became preoccupied almost exclusively by the many ways in which 'things did not join up' in the system, about the 'gaps and holes' he had to 'find a way to plug'. Listening out for the underlying topological resonance of his speech made it possible to intervene in ways that led to reconfiguring the pathways of his signifiers. In turn, this allowed the analysand to find a new way to live with his embodied knowledge of discord between the consistency and the hole.

Lacan's particular grasp of the psychoanalytic relevance of the properties of topological spaces, and his understanding of the structural significance of their nature, allowed him to clarify ways in which the body and the mind, as spaces, are neither separate nor in an easy correspondence or accord with each other. Crucial to this is the recognition that the body and the unconscious have different consistencies, between which it is only the Symbolic that can establish some link, albeit only in the form of a discontinuity.

It is in Seminar IX, *L'Identification*, that Lacan considers explicitly both the body and language as toric structures. In particular, he emphasises the division between inside and outside as misleading, given that, he argues, the two are continuously part of each other. It is the torus that interests him in particular at this point, as a topological structure, since it constitutes a spatial configuration where the inside and outside are not separated by a boundary, but are seamlessly continuous. Lacan makes a direct link between the torus and the human body, to which he presents it as homeomorphic, with the digestive tube as the hole of the torus-body (Lacan, 2020, p.344). Over a decade later, in Seminar XXII: *R S I* (1974–1975), Lacan emphasises the importance of the hole in relation to consistency,[3] the latter being what gives support to the body: 'A body, a body such as the one by which you are supported [you bear], is very precisely that something which for you has only the appearance of being that which resists, that which consists before dissolving' (Lacan, 1975, session of 18 February 1975; my translation).

It is easy to see how

> [b]oth language and the body are surfaces that combine continuity with discontinuity. Thinking of both as toric spaces makes it possible to see how the hole, the discontinuity, can never be articulated, as it is separate from the domain of the surface, but it can be circumscribed in the analytic process in a way that anchors the analysand's speech in a way that consolidates the aspect of continuity.
> (Carrington, 2020, p.164)

In terms of Figure 10.2, it is possible to see how movement between points on either side of the hole, such as from **M** to **N**, can only happen along the consistency, following its contour, rather than by traversing the void along the dotted line.

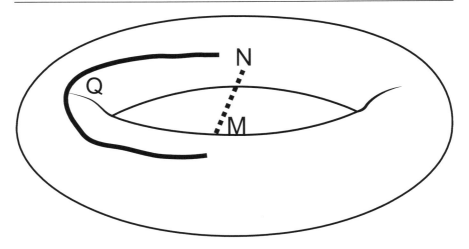

Figure 10.2 Torus pathways (1).

In the clinical encounter, the bodily experience of being – be it that of hysterical conversion or that of anorexia or addiction, either in neurosis or in psychosis, or that of bodily fragmentation in schizophrenia – enters the analytic process by means of speech. The chains of signifiers, the signifying material they constitute, build particular pathways in the space of language. These are not chains of *sens* but of *jouis-sens* (Lacan, 1973, p.517), which means that analytic interpretations in the wide sense are needed in order to facilitate access to new pathways, through acts and speech that lead to the reconfiguration of pre-established chains of signifiers.

Inasmuch as the Real is taken to denote the impossible, the real body is constituted by all that is of the body and which escapes both imaginarisation and symbolisation (Chemama and Vandermersch, 2009, p.117), that is to say, by the remainder of the encounter of the body with the other two registers of experience. A disruption at the level of the organic, the real body carries through in the realms both of the symbolic body and of the imaginary body, and it is the task of the Symbolic to re-establish some link. It fell not to further surgery but to words to find a way to do something with and around the analysand's altered state of being:

> The best that analysis can hope to offer is the chance for the analysand to work out a way to live that transforms the relationship between the hole and the consistency from an antagonism into a holed quasi-unity. This involves a movement away from trying to either escape or eliminate the interplay of these two aspects of being, towards finding a way to live with their insoluble, structural coexistence. The hole cannot be gotten rid of or escaped, it cannot be reconciled with the consistency in some kind of solid unity. The hole and the consistency are not one, and not two – they *are* only together in the relationship they have with each-other.
>
> (Carrington, 2020, p.166)

Session after session, words became the analysand's way to trace the revised contour of his bodily topology, and of the altered relationship within and between consistency and hole. The disturbances in the relationship between the two impacted on his bodily experience of compactness, in a way that emphasised the lack of accord between body and mind. This discord had always been there, as it always is, for every single subject, but for him it had become brutally unveiled. The ripples that the disruption in the body had stirred as unrest in the mind began to settle into something more bearable as soon as his attention turned to the upset he experienced about things not joining up in the world inhabited by the body rather than in the body itself: There were 'more gaps than people' in the team he had to manage, there was 'nobody' [no body] 'to ensure coherence' of the overall aims of the team, things were 'not joined up properly'. In this way, the relationship between what 'was' and what 'was not' could be talked about – about, in the sense of around.

If we take seriously Lacan's recognition of both language and the body as toric spaces, then the task of the analyst is not just that of listening, but of listening topologically to signifier pathways that trace the contours of the bodily and psychic spaces inhabited by each subject. The possibility of change arises from the reconfiguration of these pathways in ways that allow the analysand to articulate something about their desire, precisely by cutting pre-established articulations, and opening up the possibility for new connections, and therefore pathways, to emerge.

I will illustrate this operation for the case of this analysand, where a number of what I would call 'pivot signifiers' made it possible to bring together the body-mind experience in a way that reconfigured his relationship to bodily discord, permitting him to move on from the fixation with a local bodily disruption, and re-engage with the rest of his life.

These signifiers operated in the way that Freud found ambiguous words to operate, in dreams and in jokes, as links from a manifest to a latent content. He referred to these as switch points, switch words or verbal bridges. For instance, in a footnote in his 1905 paper, *Fragment of an analysis of a case of hysteria*, with reference to dreams, Freud explains:

> in a line of associations ambiguous words (or, as we may call them, 'switch words') act like points at a junction. If the points are switched across from the position in which they appear to lie in the dream, then we find ourselves on another set of rails; and along this second track run thoughts which we are in search of but which still lie concealed behind the dream.
>
> (Freud, 1905, p.65)

The role of the analyst is to hear these pivot signifiers (gap, discontinuity, joining-up, no body/nobody, failure) and to encourage the analysand to elaborate around each of them, often by doing no more than cutting the flow of his usual speech on these junctions, or by uttering one of these particular words back to him. In the space of language concatenation, these signifiers operate like the railway turntable to which Lacan refers in the session of 7 May 1969 of Seminar XVI, *D'un Autre à l'autre* (Lacan, 1969).

176 Clinical implications

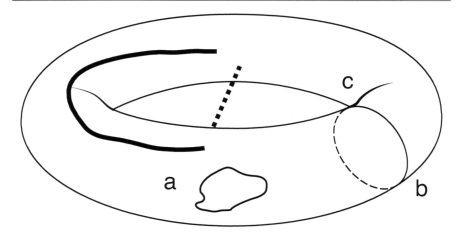

Figure 10.3 Torus pathways (2).

Keeping in mind the distinction between speech pathways that can be reduced to a point (**a**, in Figure 10.3) and those that cannot (**b** and **c**, in Figure 10.3), pivot signifiers (**P**) relate to full speech, and have a unique position at the junction of pathways such as **b**, which traces the contour of the torus 'tube', and **c**, which traces the contour of the central hole of the torus. This is consistent with Lacan's earlier view that 'a word presupposes a point with several paths through it' (Burgoyne, 2018, p.11), which is most relevant clinically as far as the possibility and mechanism of change are concerned.

Pathways **a**, **b** and **c** are all closed curves. However, note that, while the pathway labelled **a** separates the surface into two distinct regions, an inside and an outside, pathways **b** and **c** do not have that effect.[4]

In this particular clinical illustration, the articulation (in the sense of linking-up) that such signifiers made possible was not between latent and manifest content, but between speech tracing the contour of the consistency and speech tracing the hole, establishing thus some links between the spoken experience and the bodily configuration that could not be articulated directly. These signifiers operated like the railway turntable referred to (Lacan, 1969), connecting the unspoken and unspeakable bodily discord to the speakable and spoken discord at the level of experience.

The cut of a pathway such as **a** made it possible for a pathway such as **b** to emerge at the point of the cut, and for this to become linked to a pathway such as **c**. As illustrated in Figure 10.4, from talking endlessly about the stoma bag and surgery (**a**), the analysand moved to another space of speech, where he spoke with passion about the many demands and failures in the context of his relationships, especially in a work context, where he felt relatively more successful than in the field of personal interactions (**b**). Emphasising the pivot signifiers along **b** made it possible for something to be articulated about the thereto unspoken gaps and discord in the body (**c**).

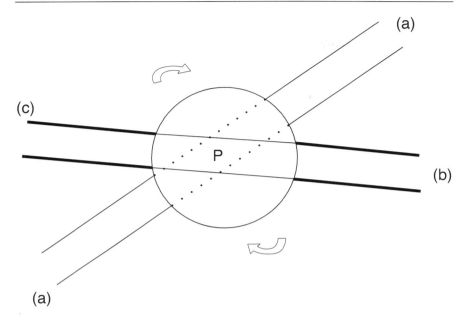

Figure 10.4 Pivot signifiers.

Some regard Lacan's later topological work as somehow separate from his earlier contributions which focused on language and the signifier. Yet this particular clinical encounter shows something about the continuity between the two stages in Lacan's formulations around the work of analysis, allowing to grasp something about the immediate, clinical relevance of topology.

Also, this particular clinical encounter brings to light, in action, the topological links between the body and language, in a sense that goes beyond the usual reference to the symbolic body, as spoken, illustrating how the Symbolic operates to establish a bodily rapport across registers (e.g. Soler, 2016). More precisely, it is the topological equivalence between the body and language that makes it possible to operate at the level of the symbolic and obtain effects across registers.

For the analysand, severe, disruptive changes in his body became something that could be talked *around* precisely because the body as space of language and the subjective experience as space are topologically equivalent.

10.3 Interpretation, the cut and the analytic act

Bion thinks of interpretation in spatial terms, as a re-mapping of a 'transformation' of what the patient presents, and the direction of travel is towards a knowing that remains 'unsaturable' (Bria, 1981, p.509) and which 'takes form as a "structurally" dramatic act' (Bria, 1981, p.510), not unlike the Lacanian cut. We prefaced the importance of the cut as analytic intervention in Chapter 6, where we acknowledged that the nature of a cut and the number of cuts (genus) have specific effects in modifying a particular spatial structure. In this sense, analytic interventions amount to acts of spatial transformation.

As Julien (1994) aptly sums it up, 'analytic interpretation is a saying that makes a knot, or else it is a bla-bla-bla with no relation to the real' (p.183). This is precisely what we can see on the torus, where interpretation creates a new pathway that brings together the consistency with the hole in the only way possible to engage desire with demand.

Vandermersch (2009) draws attention to one of the key theorems in topology, namely, Brewer's fixed point theorem, the essence of which is that every continuous transformation of a space has at least one point where this transformation has no effect (Vandermersch 2009, p.138). Given the nature of language as an open space, this theorem can only be applied once the space of language becomes closed. The operation that achieves this transformation is that of the Name-of-the-Father, without which the place of the phallus as the one fixed point would not emerge automatically (Vandermersch 2009, p.139). The implications in terms of the effects of intervention across clinical structures are important, but beyond the scope of this book.

If we take the view that psychoanalysis is less about deciphering or understanding the unconscious and giving it meaning than it is about creating the possibility of change through interventions that disrupt established unhelpful meanings manifested in the form of symptoms, then analysis works by providing an additional dimension in which movement and reconfigurations become possible. A simple illustration of this is in the move from two to three dimensions. As we saw in Figure 9.1, before a new spatial dimension becomes accessible, the two shapes are in fixed relationship to each other; they are set in a way that cannot be challenged within the confines of that space. Once a further dimension becomes available, more possibilities emerge: folding can bring the two shapes closer together or block their interaction, joining the edges can change not just the distance between the two elements, but also their relative position; twisting and joining, creating a Möbius band breaks all the parameters of the previous configuration, while freeing the elements to enter in new configurations, and produce a new flow of continuity.

Although Matte Blanco did not make explicit the connection between interventions and dimensions, the very language he employs carries spatial connotations, as he refers to lifting in the case of the repressed unconscious and to unfolding in the case of the unrepressed unconscious (Matte Blanco, 1975, p.16), both spatial manoeuvres requiring movement in a new dimension.

In its daily unfolding, clinical work is fundamentally in the nature of knot transformation. Without cutting or adding elements to the patient's story, but through disrupting long-held connections, the analyst uses the elasticity of speech to facilitate a change in the subjective position such that better use of (psychic) space becomes possible. Old pathways are disrupted; new ones become potentially available.

Lacan conceptualised this by introducing the notion of the real unconscious and the movement towards the analysis of the *parlêtre*. The course of an analysis is initially defined by the traversing of an extensive space of meaning and construction, before entering – and not every analysand chooses to go that far – into the realm of the *parlêtre*, the speaking being with its enjoying body. Access to this new domain becomes possible only after a long-drawn process aimed at the draining of jouissance. It is as if the flood of jouissance one arrives at analysis with needs to be slowly lessened, before access to another dimension becomes possible. If we think that, mathematically, the point from which various dimensions can be accessed is zero, it is a kind of equivalent psychic point that analysis moves towards, an 'origin' from which it is possible to start anew in a direction hitherto unavailable. At the same time, something structural needs to shift, akin to the radical movement from real to complex numbers, for instance. One never gets 'there' by more (or less) of the same; no amount of real numbers will form a critical mass for a complex one to emerge spontaneously. Likewise, accessing the dimension of nonsense cannot occur though adding more sense. Rather, the operation is one of subtraction of sense and of disruption, up to the point where movement in a hereto direction becomes possible.

Psychoanalysis works because the space in which we operate, through – but not exclusively in – language, is homeomorphic with the space of the Real, where changes that go beyond insight and representation can occur. The four-dimensional quality of this space is captured in the Lacanian version of the Oedipus structure, which is not triangular but quaternary, with the phallus occupying the fourth position alongside the father, mother and child (Lacan, 1956, p.461). As Bursztein (2008) clarifies, '[T]he subject of the unconscious is none other than the supposed subject of this four parties structure' (p.44).

The most common reference to a fourth dimension is to time, in addition to the three spatial dimensions that we can perceive in the world we inhabit as bodies. In the four-dimensional realm of the unconscious, the fourth dimension is not simply time, which is not to say that there is anything simple about time. Rather, it is not time in the sense that we are accustomed to.

In Seminar IX, Lacan emphasises that topologically, the nature of structural relations that constitute a surface is present at any point (Lacan, 2020, p.346). This can be seen as a very useful definition of transference – what happens at one point in the psychic space is what happens at every other point. In that sense, the need to interpret transference, as if one could do from outside it, becomes redundant, even if apparently possible. The key aspect becomes that, whatever the point at which one intervenes to modify the topology of the psychic space, that intervention will affect every point of this space. This is also what explains that in analysis the totality of the patient's life is transformed, even if in some respects the work in the consulting room can be understood as a localised intervention.

Clinical change is not something that occurs merely at the level of the Symbolic, but also in relation to the Real, in terms of shifts in enjoyment. What makes it possible for an analysand to move to a freer space within their own life is the fall of jouissance lodged in the operation of particular signifiers, and the possibility of new signifiers emerging as available for concatenation, invested with new relevance and new forms of freedom and enjoyment that are not bound to the Other.

Notes

1 Remember, from Chapter 7, the torus is structurally a product of circles, $T = S^1 x S^1$.
2 A diagnosis of psychosis was ruled out quite early on, primarily on the grounds of the analysand's particular relationship to language and to his symptom, which indicated his capacity to operate on the level of the Symbolic, even when confronted with abrupt, persistent invasions from the Real.
3 In the same seminar, Lacan specified consistency as that which exists in relation to the hole (session of 18 February 1975).
4 See note 5 in Chapter 6.

References

Bria, P. (1981). Catastrophe and Transformations. *Rivista di Psicoanalisi*, 27, pp.503–512.
Burgoyne, B. (2018). The changing forms of a research programme. In Bailly, L., Lichtenstein, D. and Bailly, S. (eds.). *The Lacan tradition: Lines of development – Evolution of theory and practice over the decades*. London: Routledge, pp.46–89.
Bursztein, J.-G. (2008). *On the difference between psychoanalysis and psychotherapy*. Paris: Nouvelles Etudes Freudiennes.
Carrington, A. (2020). The body as a spatial aspect of being. *Analysis*, (23), pp.155–169.
Charraud, N. (2004). La topologie «TBMCC» de Jaques Lacan. In Cartier, P. and Charraud, N. (eds.). *Le réel en mathématiques: Psychanalyse et mathématiques*. Paris: Agalma, pp.135–151.
Chemama, R. and Vandermersch, B. (2009). *Dictionnaire de la psychanalyse*. Paris: Larousse.
Freud, S. (1905). Fragment of an analysis of a case of hysteria, SE7, pp.1–122.
Freud, S. (1925). Negation. SE 19, pp.233–239.
Greenshields, W. (2017). *Writing the structures of the subject: Lacan and topology*. London: Palgrave Macmillan.
Hughes, T. (2013). The Klein bottle. *The letter*, 53, pp.57–85.
Izkovich, L. (2022). What is a case in psychoanalysis. *International Lacan Seminar*, 12–13 February 2022.
Julien, P. (1994). *Jacques Lacan's return to Freud: The real, the symbolic and the imaginary*. London: New York University Press.
Lacan, J. (1953). The function and field of speech and language in psychoanalysis. In Fink, B. (ed.). (2002). *Écrits*. London: WW Norton, pp.541–574.
Lacan, J. (1956). On a question prior to any possible treatment of psychosis. In Fink, B. (ed.). (2002). *Écrits*. London: WW Norton, pp.531–488.
Lacan, J. (1969). D'un Autre à l'autre, 1968–1969. Unpublished.
Lacan, J. (1973). Télévision. In *Autres écrits*, Paris: Éditions du Seuil, pp.509–545.
Lacan, J. (1975). *R S I*, 1974–1975. Unpublished.

Lacan, J. (1997 [1981]) *The seminar. Book III. The psychoses, 1955–1956*. London: WW Norton.
Lacan, J. (1998 [1975]) *The seminar. Book XX. On feminine sexuality, the limits of love and knowledge, 1972–1973*. London: WW Norton.
Lacan, J. (2020). *L'identification: Séminaire 1961–1962*. Éditions de l'Association Lacanienne Internationale. Paris: Publication hors commerce.
Matte Blanco, I. (1975). *The unconscious as infinite sets* (1975). Aylesbury: Duckworth.
Miller, J.-A. (1979). Supplément topologique à la "question préliminaire". *Lettre de l'École Freudienne de Paris*, 27, pp.127–138.
Nasio, J.-D. (1998). *Five lessons on the psychoanalytic theory of Lacan*. New York: State University of New York Press.
Nasio, J.-D. (2010). *Introduction à la topologie de Lacan*. Paris: Petit Bibliothèque Payot.
Ragland, E. (2002). The topological dimension of Lacanian optics. *Analysis*, 11, pp.115–126.
Seifert, H. and Threfall, W. (1980). *A textbook in topology*. New York: Academic Press.
Soler, C. (2016). The body in the teaching of Jacques Lacan. CFAR [online]. https://jcfar.org.uk/wp-content/uploads/2016/03/The-Body-in-the-Teaching-of-Jacques-Lacan-Colette-Soler.pdf
Vandermersch, B. (2009). Littoral ou topologie du refoulement. *La revue lacanienne*, 1(3), pp.137–144.
Vanier, C. (2015). *Premature birth: The baby, the doctor and the psychoanalyst*. London: Karnac.
Volkert, K. (2013). Projective plane; A history. *Bulletin of Manifold Atlas*, pp.1–4.

Chapter 11

Concluding comments

The question I set out to address in this book was whether thinking of the unconscious in mathematical terms could shed new light on what remains unknown in our daily experience of being. Specifically, the aim was that of exploring the unconscious as space, by addressing one specific question: can the articulation of unconscious as space, in a mathematical sense, help us learn something new about the nature of this most influential unknown in our lives?

Starting with Freud, the territory where an answer to the question of human suffering and the workings of the psyche could be sought became narrowed down to the conceptual space of the unconscious. In other words, the unconscious became recognised as the domain of the influential unknown, an unknown that puts its unmistakeable, if mysterious, mark over the human experience of finding one's place in the world, as a living being, that is to say, a being marked by its own mortality.

The focus was on the way in which psychoanalysis relates to the unknown, in particular to the unknown of the individual unconscious, and on how a spatial approach might offer some fresh insight into thinking about this unknown.

Psychoanalysis is a broad domain, its varied schools remaining united by a recognition of the place of the unconscious as a particular kind of unknown which is regarded as holding the key to understanding the origins and manifestations of human suffering. The question central to psychoanalysis was and remains aimed at what causes suffering in each and every one of us, and what determines the ways in which we respond to it. The interest is not academic, but clinical, both in origin and in orientation.

Freud's formulations around the notion of the unconscious, which constituted the core of his theoretical and clinical work, underwent developments throughout his life. What is of particular interest here is the way in which formal elements of thinking about the psyche in spatial terms underpinned the evolution of this Freudian concept. In essence, Freud abandoned early attempts to identify specific locations for mental processes in favour of considering the relative positions of the agencies involved in registering and processing the experience of being.

Only a small number of subsequent analysts engaged explicitly with the question of the relevance of space to understanding the nature and operation of the unconscious, and the particular kind of knowledge that it constitutes. Up to this point, these contributions have not come together in a cohesive body of work.

Both Freud and Lacan emphasised the discontinuity and the lack of symmetry between conscious and unconscious processes and aspects of being. In Freudian terms, the unconscious is not a kind of consciousness of which we are not aware yet, it is not the 'sub-conscious' of every day language, it is not just a 'yet unknown' entity, but a kind of unknown that is not directly knowable. We have explored this formally in terms of possible ways of understanding the unconscious as a spatial structure.

The Freudian unconscious is an 'elsewhere' of a particular kind, a spatial unknown that cannot be either accessed directly or ever apprehended in its totality. This view changes the common notions of *internal* and *external* and puts on new foundations the questions of the dis/continuity to which the concept itself was introduced by Freud as a necessary hypothesis.

In this sense, it is important to remember that this book uses a series of mathematical constructions to examine a possible truth, and all it can hope to offer is a possible new reading: it is not a truth that the unconscious IS a four-dimensional space, but rather an examination of what thinking of it in this terms has to offer. Likewise, in terms of topological understanding.

Analysis follows a course that begins by traversing a space of meaning and construction, before approaching a new kind of territory, that of the unknowable unknown, with a movement that has a structure-modifying properties. It is abrupt, even if not sudden. Accessing this other kind of space, beyond meaning, does not happen purely by more of the same, by producing more and more meaning. Instead, the prerequisite is that of draining, as if only once enough sense has been removed can the gateway to this realm of the *parlêtre* be accessed. This is akin to movement to another dimension, through the origin set at zero, wherefrom a new dimension opens up from 'nothing'. In other words, the operation of analysis upon the subject is one of transmutation.

In essence, the main thesis put forward here is that the unconscious is structured like a space, operating in ways defining of the interplay between spatial dimensions. Approaching it as a spatial structure yields new insights into the working of the mind and of what is at stake in our clinical work, what we intervene upon, how and to what effect.

In order to develop an understanding of the unconscious as space, it is essential to consider the notion of dimension. This is a term often used in daily life, often metaphorically, but which has its specific and complex signification, mathematically, which carries particular relevance to our work.

Whether it can be perceived by the senses or not, any space is defined by its dimensionality. The argument constructed here is that the unconscious can be thought of as a four-dimensional space, which means that, by necessity, it can never be apprehended in its totality, but only in a seemingly fragmented way, through the means of perception we have available in what we experience as three-dimensional physical space. Mathematically, a four-dimensional space includes the entirety of any three-dimensional space, and something besides. What comes across as fragmented and repetitive can be understood as the necessary consequence of dimensional incompatibility between our capacity to perceive and experience, and the structure of that which is to be perceived, given that something four-dimensional would not fit into three-dimensional space without fragmentation, overlap or repetition.

In both Freudian and Lacanian terms, repetition is the manifestation of the unconscious and is present in every subject. Its universality is indicative of a structural underpinning. The inevitability thus at play is often seen as an automatism that evades the intention of the subject. The main implication of regarding this as a necessary consequence of spatial structure is that repetition is no longer to be viewed as a symptom, but as a consequence of bringing into experience one single, unitary inscription lodged in a space of higher dimensionality. In other words, it cannot he helped.

What is traditionally regarded as the infinite nature of the unconscious could therefore be thought about in a different way: rather than a domain akin to an interminable sequence of elements, the unconscious can be conceptualised as a domain whose totality can never be apprehended not because of its vast size, not because there is not enough time to do so, but because its totality resides in another dimension and is therefore structurally out of reach.

What I submit here for consideration is not the idea that the unconscious is located *in* some impenetrable hyperspace, but that its structure is that of a four-dimensional space. More specifically, I focus on the relevance of the interplay between three- and four-dimensional spaces, namely, on the absence of continuity and coherence at the level of experience, which can be understood as the experience of four-dimensional processes which are collapsed or reduced to the habitual perception of three-dimensionality, and thus unavoidably fragmented and seemingly disjointed.

In this sense, we could say that the so-called 'internal' world is four-dimensional, while the perceived external world is three-dimensional and the body makes a sort of cut between the two.

What seems to be a sequence of recurring experiences, made chronological by speech, could be more coherently understood as parts of the same one unconscious construction – the fundamental phantasy. Phantasy, in turn, is in a direct relation to the Real; it is the veil that masks the prohibition. The construction of the fundamental phantasy is one which can be reduced to a minimal axiom, which is impersonal. The Lacanian view of analysis is one where the subject traverses it, to move into a position where new phantasies can arise. The term 'traversing' has strong spatial connotations, which raises the question of what may be the topology of the pass.

If we understand the difficulty with reaching the unconscious as such, in a way that could produce direct change, as a question of inaccessibility, our approach would be to find a better route to the 'far away place', a kind of shortcut. If, however, we recognise this same difficulty as one produced by structural impossibility (in carrying over any entity from a space of higher dimensionality to a space of lower dimensionality), then the work of analysis takes an entirely different orientation, as it accompanies the subject on a journey of creating a way of being with impossibility itself, rather than blaming themselves for it or trying to resolve it.

The same is true about understanding this space as a topology marked by discontinuity, where possibility amounts to movement in a landscape marked by impossibility. Satisfaction is not elusive merely because it could be found only in a space barred by a 'no entry' sign, but rather, because where we expect to find it, there is nothing but a hole. We try to fill this hole with phantasy and love objects, we select something that might satisfy the drive, yet – without fail – the chosen object fails, everything fails. Complete satisfaction is not merely forbidden or out of reach, but fundamentally and structurally impossible. The failure is necessary, yet it operates as an engine of search, tracing in this way the path of desire and therefore of life itself.

Index

Note: Italic page numbers refer to figures and page numbers followed by "n" denote endnotes.

Abbott, Edwin A. 67, 69, 77n4, 81; *Flatland: A Romance of Many Dimensions* 67–71, 81, 89
Adams, Douglas: *Hitchhiker's Guide to the Galaxy* 46
affecttrauma model 9, 10
algebraic topology 116
The Ambassadors (Holbein) 95n7
analysis situ 102, 141
analytic act 50, 177
asymmetric logic 21

Bion, Wilfred Ruprecht 19–20, 23, 88
Blacklock, M. 30, 70, 77, 77n2, 140
Blum, V. 109, 162
The Book of Sand (Borges) 43
border 108, 118, 148
Borges, J.L. 43; *The Book of Sand* 43
Borromean knot 30, 57–58, 105, 109, 138n4, 141, 145, 147–148, 149, 151n2, 155, 157
boundary 10, 37, 40n4, 47, 66, 73, 84, 99, 104, 105, 116–118, 120, 122–124, 128, 130, 131, 138n7, 146, 148, 149, 157, 163, 169, 170, 173
brane worlds 149
Burgoyne, B. 165
Bursztein, JeanGerard 12, 18, 30, 57, 58, 88, 95n5, 105, 115, 145–146, 155, 157, 179

calculus 19–20, 54, 102
Cantor, Georg 44–45, 48, 83, 93–94, 102
Carrington, A. 171
Charraud, N. 145, 166
Chemama, R. 54, 89

Civilisation and Its Discontents (Freud) 82, 145
classification theorem 117
Cléro, J.P. 4, 22, 46, 90, 109, 149
clinical implications 165–180; analytic act 177–180; clinical illustration 171–177; interpretation 177–180; topology in clinic 165–171
closed surface, classification of 117
closed twodimensional manifolds 117
Coetzee, John Maxwell 132; *The Life and Times of Michael K* 132
CohenTannoudji, G. 5
complex numbers 48–51, 56–57, 59, 73, 86, 146, 179
condensation, and unconscious 36
conscious (Cs) 10
continuity 115, 156–157; defined as 115; in higher dimension *159*
continuous space 15, 38, 50
cross-cap 120, 123–131, 134, 135, 137–138n4, 138n5, 162, 163, 166, 169, 170
crumpled space 38, 161
the cut 131–132, 138n8, 155

Dedekind, Richard 44
determinism 6
dimensions: in the clinic 90–92; higher 73–75; lower 73–75
displacement, and unconscious 36
Eddington, Sir Arthur Stanley 24
edge 15, 37, 64, 73, 83, 84, 109, 117, 118, 120–12, 128, 131, 134, 138n8, 149, 169, 170, 178
Eigen, Michael 24, 43, 47

Einstein, A. 84–85, 150; relativity theories 150
embedded torus *120,* 122
Euclid 64, 84
Euclidean geometry 6, 29, 37, 48, 84, 134; threedimensional 38; undefined terms in 83
Euclidean lines 127
Euclidean plane 127
Euclidean space 6, 14, 38, 64, 134
Euclidean threedimensional space 128, 141
exclusivity of fourth dimension 79–94

Fink, Bruce 37
Flatland: A Romance of Many Dimensions (Abbott) 67–71, 81, 89
Flatterland: Like Flatland, Only More So (Stewart) 71, 117
fourdimensional space 87–90, 135; as cultural object 140; knot in 142
The four fundamental concepts of psychoanalysis (Lacan) 95n7
fourth dimension; exclusivity of 79–94; unconscious as inaccessible 79–94
Fragment of an analysis of a case of hysteria (Freud) 175
Freud, S. 3–4, 64, 74, 79, 81, 82, 86, 101, 105–107, 116, 131–132, 135–137, 156, 159–160, 169, 182, 183–184; archaeological references 161; *Bejahung* 111; *Civilisation and Its Discontents* 145; first topography 16–17; *Fragment of an analysis of a case of hysteria* 175; *Inhibitions, Symptoms and Anxiety* 145; *The Interpretation of Dreams* 9, 13, 35, 150; maps 12–18; *Project for a Scientific Psychology* 4–5, 9, 13, 35, 79, 159; Rat Man case 109; second topography *16, 17;* on space 5, 14; *Studies on Hysteria* 13; tangle of dreamthoughts 39; on unconscious 6, 8–9; *The Unconscious* 10; *Unerkannt* 105
Freudian unconscious 4, 6, 8, 41; and consciousness 53; hypothesis of 86
The Function and Field of Speech and Language in Psychoanalysis 167
fundamental polygon 118: Klein bottle 123, *123*; Möbius band in *119*; projective plane/crosscap 124, *127*; torus in 121, *121*

Garella, A. 25–26
Greenshields, W. 107, 115, 120, 121, 140, 167
halfdisklike neighbourhood 116, 118, 120
Hawking, Stephen 66, 77n5, 149
Hermann, Imre 18, 28, 156
higher dimensions 71–75, 137
Hinton, Charles 64, 73–74, 77n2, 158
Hippasus 55
Hitchhiker's Guide to the Galaxy (Adams) 46
holes: defined as 133; as manifestations of radical discontinuity 105; topology 132–134; and the unconscious 104–106
homeomorphic spaces 60n6, 155
homeomorphic transformations 115
homeomorphism 135, 142
hyperspace 14, 66, 83–87
hypokeimenon 140
imaginary 23, 29, 30, 31n11, 42, 50, 51–53, 56, 57, 71, 72, 82, 90–93, 105, 109, 118, 126, 145, 146, 149, 157, 158, 161–163, 166, 174
imaginary numbers 42, 48, 50–51, 59, 71, 85, 170
impossible/impossibility: holes and the unconscious 104–106; mathematical representations of 102–104; negations of 110–112; no dimensions and inscription of 93–94; Oedipus complex as prohibition veiling impossibility 106–110; structures of 101–112
incompleteness 41–60; in the unconscious 53–58
infinity 42–43, *43*; described 41; essence of 41; as recurrence 47; unconscious as 41–60
Inhibitions, Symptoms and Anxiety (Freud) 145
instinct (drive) 9
interpretation 19, 20, 25, 39, 59, 85, 89, 105, 133, 147, 148, 161, 167, 174, 177–180
The Interpretation of Dreams (Freud) 9, 13, 35, 87, 150
intervention 45, 50, 105, 111, 140, 145, 147, 153, 155, 156, 167, 171, 172, 177–179
IPA (International Psychoanalytical Association) 16, 81, 91, 106, 160–161, 165–166
irrational numbers 57–58
isotopy 40n5

Jordan curve 112n5
jouissance 111, 148, 155, 157, 162, 163, 170, 179, 180
KaluzaKlein theory 86
Kasner, E. 5, 23, 31n10, 45, 64, 65, 74, 94, 118
Klein, Felix 20, 142
Klein, Melanie 91
Klein bottle 123–124, 137n3; fundamental polygon representation 123, *123*; Möbius bands and 124, *125*; selfintersection 135; in threedimensional space *126*
knot(s) 31n12, 141–145; Borromean 30, 57–58; in clinic 146–148; and dimensions 148–151; in fourdimensional space 142; in mathematics 141; as structure and pathways through language 145–146; in threedimensional space 142; trefoil 142; unconscious as 30, 140–151
knot invariants 143
knottedness 142, 149
knot theory 22, 29, 39, 140–141
knowledge 24–25; Bion on types of 19; in four dimensions *25*; and mathematics 25; unconscious as 19–25

Lacan, Jacques 3–4, 6, 15, 18, 22–23, 46, 74, 75, 76, 81–82, 87–91, 95n7, 105–109, 115, 116, 119–120, 122–124, 127–128, 130–135, 137–138n4, 138n8, 140, 145, 148, 168–170, 173, 183, 184; about embeddedness 137; about interior eight 146; affinity of holes 133; critique of IPA/Kleinian approach 161; elaboration of the Symbolic 54; *The four fundamental concepts of psychoanalysis* 95n7; *La troisième* 155; *L'identification (1951–1952)* 26; link between psychoanalysis and topology 132; Lschema 161–162; optical schema in seminar 161; *The position of the unconscious* 109; psychoanalytic concepts 131; psychoanalytic theorising 30; Rschema 162; seminar on *Identification* 135; spatial thinking 29; spatial view of unconscious 156; topology for 30; *traumatisme* (trauma), by 134; unary trait 55; on unconscious 23, 36–37; use of complex numbers 56; use of mathematics 166

'lacking in lack' notion of the Real 59
Laplanche, J. 10, 12, 35
La troisième (Lacan) 155
Levinas, E. 4
L'identification (1951–1952) (Lacan) 26
The Life and Times of Michael K (Coetzee) 132
Listing, Benedict 116, 118
lower dimensions 71–75
Lschema 161–162

Magee, B. 3–4, 137
Maimonides 18
Mandelbrot, Benoit 66
Marcus, S. 48, 59, 101
mathematical representations of impossibility 102–104
mathematics: Balibar on use of 23; Lacan's thinking 22; and psychoanalysis 18–19, 22–23
Matte Blanco, Ignacio 18–23, 68, 76, 81, 165, 178; on mental space 26–27; on space 26–27; on space in mathematical sense 28; spatial description of dreams 28; *The Unconscious as Infinite Sets* 21
Miller, J.A. 110, 133, 162, 170
Minkowski spacetime 85
miseàplat 143, 148, 157, 162
Mlodinow, L. 66, 77n5
Mnem (mnemic system) 14
mnemic traces 22
Möbian structures 95n5
Möbius, August Ferdinand 20, 64, 118; *On higher space* 64
Möbius band (MB) 15, 117–120, 128, 146; closed *126*; crosscap into 131; embedded in three dimensions *15*; in fundamental polygon representation *15*, *119*; Klein bottle and 124, *125*; as topology 120; and torus 122
Möbius strip 82
Moncayo, R. 50, 57
Moore, Allan 70; *Tales of the Uncanny* 70
Nasio, J.D. 128, 130–131, 170–171
negations of impossibility 110–112
n'espace 148
Newman, J. 5, 23, 31n10, 45, 64, 65, 74, 94, 118
New Realism 79
Newton, I. 41, 150
no dimensions 68; and inscription of impossibility 93–94

nonEuclidean geometry 64
number line 43–44
number theory 53

Oedipus complex 101, 106, 169; as prohibition veiling impossibility 106–110
onedimensional space 42
onesided surfaces 118
On higher space (Möbius) 64
open space boundary 37, 117, *117*
orientability 119
overdetermination, and unconscious 36

parlêtre 122
Paty, M. 85–86
Peirce, Benjamin 53
Penrose, Roger 149
*percipien*s 158
phallus 55, 57, 101, 111, 127, 146, 170
physical reality 149
pivot signifiers 175–176, *177*
Plank, Max 150
Plato 56, 72
pleasureunpleasure principle 10
Poincaré, H. 64, 86, 131, 141; *Analysis Situs* 141
Pontalis, J.B. 10, 12, 35
The position of the unconscious (Lacan) 109
postFreudian emphasis 19–30
preconscious (*Pcs*) 10, 35–36
primary mental processes 11
prohibition veiling impossibility, Oedipus complex as 106–110
Project for a Scientific Psychology (Freud) 4–5, 9, 13, 35, 79, 159
projections 39, 73
projective plane/crosscap 124–131, 135, 137–138n4, 138n6, 146, 169, 170; alternative representation *130*; decomposing *129*; fundamental polygon representation 124, *127*
proportionality 46–47
psyche, in spatial terms 8
psychic agencies: relative positions of 15; spatial approach 16; topology of *14*
psychic locality 14
psychic space 25
psychic structures 156–163; development of 157
psychoanalysis 115, 131, 153, 179, 182; cause of suffering 11; in clinic 163; concept of projection 160; and determinism 6; effects of 165; and mathematics 18–19, 22–23; spatial dimensions in 75–77; and topology 132, 156; and unknown 5
'psychotopology' 109

quantum mechanics 150–151
quantum theory 150

Ramanujan, Srinivasa 45
rational numbers 44
real 4, 11, 20, 23, 29, 30, 31n11, 37, 38, 42, 44, 48, 50–54, 56–59, 72, 75, 81, 85, 89, 90, 102, 107, 109, 124, 134, 145, 146, 149, 155–160, 162, 166, 168, 172, 174, 178, 179, 184
real number 51
recurrence 45; infinity as 47; spatial dimensions of 63–76
regression 10
Reidemeister, Kurt 143
Reidemeister moves 143, *144*, 147
repetition 73–75
repression 9–11, 21
Republic (Plato) 72
Riemann, D. 84
Riemann sphere 112n4
Romanowicz, M. 50, 57
Rosolato, G. 99, 105, 111
Rosset, J.P. 54
Rschema 162

Sautoy, Marcus du 4, 99
Schläfli, Ludwig 72
secondary mental processes 11
Secor, A. 109, 162
shadows: and recurrences 63–76; spatial dimensions of 63–76
sidedness 118–119
singularities 86, 95n6, 102, 112n1
space 38; basic notion of 6; continuous 15, 38, 50; crumpled 38, 161; defined as 5; dimensionality to 185; Euclidean 64, 134; fourdimensional 87–90; Freud on 5, 14; homeomorphic 60n6, 155; Matte Blanco on 26–27; psychic 25; topological 103–104; unconscious as 6, 25–30
spatial dimensions: flatland and beyond 66–71; from higher to lower dimensions: projection 71–73; lower to higher dimensions: repetition 73–75; rigour of

63–76; of shadows and recurrences 63–76; spatial dimensions in psychoanalysis 75–77
spatiality 20; implied, of unconscious 25
spatial unconscious 156–163; clinical dimensions 161–163; embodied relationship 161; identification and 160–161; recurrence of suffering 157–160; subjectivity and 160–161
Sphere passing through linear space *69*
Stewart, I. 41, 43, 71, 143; *Flatterland: Like Flatland, Only More So* 71, 117
structures of the impossible 101–112
Studies on Hysteria (Freud) 13
symbolic 20, 22, 23, 30, 31n11, 51, 53–59, 71, 72, 75, 81–83, 85, 86, 89–91, 93, 101, 105, 107, 109, 111, 118, 122, 126, 145, 146, 149, 157, 159–163, 166, 172–174, 177, 179, 180n2
symmetrical logic 21

Tait, Peter 140
theory of relativity 84–85
threedimensional Euclidean space 116
threedimensional space 118–119; embedding surface into 122, 134–135; knot in 142; trefoil knot in 142
Tibetan Buddhism 79–80
topography 9, 10–14, *16*, *17*, 25, 27, 29, 105, 132, 166
topological spaces 40n2, 103–104
topological surfaces 117
topology 18–19, 22, 40n2; algebraic 116; branches of 116; cartographic illustration of structure 115; in clinic 165–171; concepts of 116; the cut 131–132; defined as 116; embeddedness 134–137; Euclidean geometry 37; holes 132–134; hyperspace 14; Klein bottle 123–124; and knot theory 29; for Lacan 30; in mathematical sense 23; Möbius band 117–120, *119*; primer 116–132; projective plane/crosscap 124–131; of psychic agencies *14*; and psychic phenomena 29; and space 29; torus *120,* 120–123, *121, 122*; of twodimensional manifolds 115; unconscious as 115–137
torus 120–123, *174, 176*; demand and desire on *122,* 122–123; embedded *120,* 122; in fundamental polygon representation 121, *121*; and Möbius band 122
transformations 115; homeomorphic 115
trefoil knot 142
Tucker, A.W. 103
twodimensional space 48–50, 116, 158; threedimensional appearance of 134
twodimensional surface 137–138n4
unary trait 54, 55
unconscious 10, 35–36; condensation 36; displacement 36; and exclusivity of fourth dimension 79–94; as fourdimensional space 87–90; Freud on 6, 8–9, 13; holes and 104–106; as inaccessible 79–94; as inaccessible space 35–39; as inaccessible space between points of encounter 58–60; as infinity 41–60; as knots 140–151; as knowledge 19–25; in mathematical terms 3; as nonEuclidean topological space 6; overdetermination 36; in psychoanalytic sense 3; as space 6, 25–30; as spatial structure 15; symmetrical logic 21; as a system 12; as topological space 115–137
The Unconscious (Freud) 10
The Unconscious as Infinite Sets (Matte Blanco) 21

unknown 24; influential 9; and psychoanalysis 5

Vandermersch, B. 54, 89, 105, 127, 146, 147, 178
Vanheule, S. 110, 133, 162
Verhaeghe, Paul 23–24, 30, 107, 155
Vivier, L. 93, 104

Yalom, I.D. 105, 134

Zeeman, E.C. 149

For Product Safety Concerns and Information please contact our EU representative GPSR@taylorandfrancis.com Taylor & Francis Verlag GmbH, Kaufingerstraße 24, 80331 München, Germany

Printed and bound by CPI Group (UK) Ltd, Croydon, CR0 4YY
11/06/2025
01899244-0001